T0138927

A Microscopic Submarine
in My Blood

A Microscopic Submarine in My Blood

Science Based on *Fantastic Voyage*

How technology has transformed into reality a classic
Hollywood science-fiction movie about a submarine crew that
shrinks to microscopic size and ventures into the body

Sylvain Martel

PAN STANFORD PUBLISHING

Published by

Pan Stanford Publishing Pte. Ltd.
Penthouse Level, Suntec Tower 3
8 Temasek Boulevard
Singapore 038988

Email: editorial@panstanford.com
Web: www.panstanford.com

British Library Cataloguing-in-Publication Data
A catalogue record for this book is available from the British Library.

A Microscopic Submarine in My Blood: Science Based on *Fantastic Voyage*

ISBN 978-981-4745-78-9 (Hardcover)
ISBN 978-981-4745-79-6 (eBook)

Printed in the USA

To my father, a victim of cancer

Logic will get you from A to B. Imagination will take you everywhere.

Albert Einstein

The true sign of intelligence is not knowledge but imagination.

Albert Einstein

To raise new questions, new possibilities, to regard old problems from a new angle, requires creative imagination and marks real advance in science.

Albert Einstein

Contents

Preface   xiii

How to Read This Book   xvii

Introduction   1

1 The Location   5
   1.1 Landing at Los Angeles Airport
      (Scene 00:00:48–00:01:05)   6
   1.2 Transiting from the Airport to the Facility by Car
      (Scene 00:03:05–00:04:03)   6
   1.3 Secret Location of the Facility
      (Scene 00:05:48–00:06:58)   7
   1.4 Entering the Underground Level
      (Scene 00:07:00–00:07:15)   8
   1.5 Transiting through the Underground Floor and
      Going to the Upper Floor (Scene 00:07:22–00:07:57)   9
   1.6 Mechanical Stairs and Adjacent Facilities
      (Scene 00:07:45–00:07:52)   11
   1.7 Security Desk at the Entrance
      (Scene 00:07:52–00:08:00)   14
   1.8 Identification Check (Scene 00:08:00–00:08:17)   14
   1.9 Transit towards the Facility (Scene 00:08:18–00:08:38) 15
   1.10 Signs and Security at the Entrance of the Facility
      (Scene 00:08:38)   16

2 The Interventional Facility   17
   2.1 The Preparation Room (Scene 00:08:38–00:09:17)   17
   2.2 First Look at the Interventional Facility
      (Scene 00:09:17–00:10:50)   19

2.3 Name of the Facility (Scene 00:10:50–00:11:27)   21

2.4 Electronic Cabinets (Scene 00:11:48–00:12:00)   23

3 The Team at the Planning Briefing   **25**

3.1 Maximum Time to Accomplish the Mission
(Scene 00:11:35–00:11:38)   25

3.2 Goal of the Mission (Scene 00:12:06–00:12:14)   26

3.3 A Woman in the Team (Scene 00:12:40–00:13:10)   27

3.4 Team Members (Scene 00:13:10–00:13:44)   27

3.5 Beginning of the Briefing (Scene 00:13:52–00:14:23)   29

3.6 Blood Flow Rate and Its Potential Impact on
the Miniature Submarine (Scene 00:14:23–00:14:54)   33

3.7 Recovery Route (Scene 00:14:54–00:15:26)   34

3.8 Tracking the Position of the Submarine
(Scene 00:15:30–00:15:46)   34

3.9 Danger of Attack (Scene 00:15:56–00:16:18)   35

3.10 Sense of Humor (Scene 00:16:20–00:16:30)   35

4 The Submarine   **37**

4.1 External View of the Submarine
(Scene 00:18:31–00:18:35)   37

   4.1.1 Crewless Submarines   37

   4.1.2 Crewed Submarines   39

4.2 Entering the Submarine (Scene 00:18:35–00:19:18)   41

4.3 Propulsion Power (Scene 00:19:46–00:20:45)   41

   4.3.1 Microscopic Particles to Power the Submarines   42

   4.3.2 Superconducting Power   42

   4.3.3 Ferromagnetic Propulsion Engines   43

   4.3.4 Superparamagnetic Propulsion Engines   46

   4.3.5 Iron–Cobalt versus Iron Oxide Propulsion
Engines   49

   4.3.6 Generation of a Magnetic Field Sufficiently
High to Bring the Propulsion Engines of the
Submarine at Full Regime   50

4.4 At the Pilot Seat (Scene 00:21:37–00:21:52)   52

4.5 Onboard Weapon (Scene 00:21:52–00:23:20)   54

4.6 Name of the Submarine (Scene 00:24:50–00:25:00)   55

5 Miniaturization   **57**

5.1 Phase 1 of the Miniaturization Process (Scene 00:25:05–00:28:10)   57

5.2 Phases 2 and 3 of the Miniaturization Process (Scene 00:28:10–00:34:38)   63

6 Injection   **69**

6.1 Beginning of Injection with a Syringe (Scene 00:37:09–00:37:41)   69

     6.1.1 Injection Being Done Close to the Target   69

     6.1.2 Injection Being Done Far from the Target   73

6.2 Velocity of the Submarine during the Injection (Scene 00:37:41–00:38:12)   75

7 Traveling in the Artery   **81**

7.1 Size of *Proteus* (Scene 00:39:25–00:39:35)   81

7.2 One Hundred Thousand Miles Long (Scene 00:40:22–00:40:24)   82

7.3 Speed of the Submarine (Scene 00:41:29–00:41:32)   83

7.4 Distance to the Next Branching Artery (Scene 00:41:32–00:41:35)   84

7.5 Submarine not Responding Due to a Strong Blood Flow (Scene 00:41:35–00:43:25)   85

8 Tracking the Position of the Submarine   **87**

8.1 Position of the Submarine Being Indicated on a Large Map (Scene 00:45:30–00:45:58)   87

8.2 The Tracking System (Scene 00:46:00–00:46:36)   92

     8.2.1 Tracking with the Technology as in the Movie (PET-Based Tracking)   92

     8.2.2 MRI vs. Ultrasound, X-Ray, and OCT-Based Tracking   94

     8.2.3 Fundamental Principle of Detecting the Location of a Microscopic Submarine in the Body   94

8.3 The Tracking Position of the Submarine Shown on the Display (Scene 00:46:47–00:47:00)   99

9   Propulsion and Steering   **101**
9.1   Go with the Flow (Scene 00:47:27–00:47:30)   101
9.2   Pulsatile Flow (Scene 00:49:37–00:50:04)   104
9.3   Propelling and Steering Systems
       (Scene 00:50:13–00:50:16)   105
       9.3.1   Microjets   105
       9.3.2   Magnetic Resonance Navigation   106
       9.3.3   Dipole Field Navigation   109
       9.3.4   Fringe Field Navigation   112

10   Navigating in the Bloodstream   **113**
10.1   Navigating in Larger Blood Vessels
        (Scene 00:54:00–00:54:14)   113
10.2   Navigating in Narrower Blood Vessels
        (Scene 00:54:14–00:54:42)   121

11   The Crew   **123**
11.1   Low in Oxygen (Scene 00:56:00–01:06:56)   123
11.2   In the Human Mind (Scene 01:27:11–01:28:26)   123
11.3   Crew Members Exiting the Submarine
        (Scene 01:30:06–01:30:15)   124
11.4   Crew Members Swimming towards the Target
        (Scene 01:30:48–01:31:07)   125
11.5   Armed Crew Member (Scene 01:31:16–01:31:30)   127
11.6   Destroying the Target by Pointing and Shooting
        the Laser to the Right Locations
        (Scene 01:31:30–01:32:32)   128
11.7   *Proteus* Being Attacked by the Body Defense
        System (Scene 01:34:09–01:35:47)   129

Conclusion   131

Index   133

Preface

When I was a young boy, I saw through the screen of our family television set a fiction adventure movie that I would never forget. The title of the movie was *Fantastic Voyage*. It was the story of a submarine with a few crew members being shrunken to microscopic size before being injected in the bloodstream of a scientist. The ultimate goal was to reach the brain to repair the damage caused by a brain injury.

What I did not realize at the time is that, later, I would become a scientist and researcher myself, and the findings of my research would lead to the implementation of the first medical interventional facility enabling operations based on a scenario similar to the one that was presented in this well-known classic science-fiction movie. What is also amazing is the fact that this first and unique infrastructure would be inaugurated August 24, 2016—exactly 50 years after the release of this film in August 24, 1966. Realizing this exceptional opportunity, I thought that it would be interesting to share what have been parts of a "roller-coaster" but also an exhilarating research adventure for me and my team members. More specifically, I would describe in a simple manner accessible to all, how the vision of Hollywood has been translated and made possible by the technological advances that were successfully integrated half a century later in this first-of-a-kind medical interventional facility. The outcome of that is this book.

This infrastructure is really the result of my own adventure with graduate students and colleagues who helped me realize it through 15 years of intensive and extremely risky research and development efforts. Such an adventure really began in 2001, when I gradually moved from the Massachusetts Institute of Technology (MIT) to Polytechnique Montréal to establish a new research laboratory. It was named the NanoRobotics Laboratory. I did not select the name; the University did, being inspired by one of my presentations combining robotics and nanotechnology, and the fact that anything related to nanotechnology was particularly popular during this period. I made such a transition to Polytechnique Montréal for a position as a university professor, believing that I could make an important impact on our society by gaining the authority to decide on my overall research strategy.

The problem was that to make such an impact, I needed a good idea to start with. It is by restricting my thoughts to the medical field while keeping in mind the name of the lab that I finally got the idea that would define the new research path for many years to come. Robotics enables the displacement control of objects along a predefined trajectory, while *nano* suggests very small objects. By combining both, you get NanoRobotics. This naturally led my thoughts towards the new goal of the lab: the development of methods to navigate small objects in the bloodstream. I chose the bloodstream because there are close to 100,000 km of blood vessels in each human adult. More specifically, this vascular route, with a length equivalent to two and a half times the circumference of planet Earth at the equator, would allow small objects to reach any parts of the body following navigation along a predefined path.

But to seek research funding, I needed to define first the how and the why. The latter is related to the motivation to fund the project, where the potential impact generally plays an important role in the decision process. I suspected, and most people agreed with me, that no one would fund me if I would attempt to navigate objects in people's bloodstream without a very, or I must say extremely, good reason. Curing a cold or treating a headache certainly did not qualify as an extremely good reason. I need to find an application where people would not have the choice but to inject these navigable objects in the bloodstream. A first obvious medical application was cancer therapy.

Indeed, actual data suggest that about half of men and one-third of women in developed countries will be diagnosed with cancer at some point during their lifetime. Although many therapeutic agents have been synthesized and huge research efforts have been put forward over the last decades, the final result is that one person still dies of cancer every 5 seconds. With an excess of 14 million cancer patients and more than 8 million cancer-related deaths per year worldwide, there was certainly a need for a disruptive new approach. Since the main problem is that therapeutic agents do not generally reach their target, an obvious strategy made possible by combining robotics with nanotechnology was to mimic the approach suggested by this Hollywood movie by using some sort of tiny submarines with their armed crew members being navigated in the bloodstream to the site of treatment, and that would include, as seen in the movie, deep regions of the brain. As for the "how," what we proposed appeared so fictional and unbelievable that, initially, funding was unsuccessful for a while. But the many years of perseverance and the convincing results that came later have led to a modern concretization of the movie *Fantastic Voyage*. So, to find out the "how," simply read the following pages. I believe that you will be more amazed by the reality than by the fiction.

Sylvain Martel

How to Read This Book

For a better reading experience, it is highly encouraged to read this book while watching the movie. I refer to this new book format as a "moviebook." First, you must get the original version of the 1966 movie *Fantastic Voyage*, written by Harry Kleiner.

In each chapter, there will be sections where each section will have a title related to a particular scene of the movie. The title of the scene will be followed by the starting time of the scene. This starting time corresponds to the total elapsed time from the beginning of the movie. The starting time will be followed by an ending time. Similarly, the ending time indicates not only the end of the scene but also the total elapsed time from the beginning of the movie.

For example, in Chapter 1, Section 1.1 has the following title: "Landing at Los Angeles Airport (Scene 00:00:48–00:01:05)." This indicates that this particular scene begins 48 seconds after the beginning of the movie and ends 1 minute and 5 seconds after the beginning of the movie.

The title of the scene will be followed by a short description of the corresponding scene. You can fast-forward the movie to the next scene, or you can enjoy watching the movie and temporarily stop at each scene. It is better to watch the scene first before reading the text in the corresponding section before resuming watching the movie until the next scene.

The description of the scene at the beginning of each section should match the scene of the movie. If not, you may have to apply a small offset from the elapsed time indicated at the beginning of each section in order to calibrate the time.

To calibrate the time, if necessary, you must do the following. When you see the plane touching ground for the first time, the elapsed time from the beginning of the movie should be 1 min 05 s (or very close). If necessary, your time must be adjusted accordingly.

Have a good watching and reading.

Introduction

THE MAIN GOAL of this book is to introduce the reader to various technologies that have been used to implement the first medical facility that mimics the scenario of the movie *Fantastic Voyage*. This book not only gives a glimpse of how technologies have evolved since the release of this movie 50 years ago, but also shows, in a humoristic way, some changes in social comportments.

To do just that, a personal tone is used throughout the book. The technological content is also written at the introductory level in order to show the basic principles without going in too many technical details. There are no complex equations in this book. Indeed, with all the sciences and technologies involved in the making of this new medical facility, by not restricting the content to the fundamental ideas and principles, the book could have easily increased to thousands of pages filled with complex mathematical equations and chemical compounds, enough to scare any scientist.

The fundamental motivation behind this book is to show to the younger generation (but also to the older ones) that sciences and engineering can be exciting. Indeed, engineers today do more than building roads and bridges; they also put in place the next generation of medical platforms and technologies, and the following chapters are written with facts and images that aim at showing just that.

In fact, with the role of technologies that is increasing at a very fast pace every day in all sectors, including medicine, I would not be surprised to see, in a not too far future, engineering becoming a specialty in medicine. Considering that the technology being presented in this book, which was conceived with engineers with the help of scientists and medical specialists, could save millions of lives in a relatively short time, the impact of the relatively small group of engineers and scientists who were involved in the implementation of these new platforms could be seen differently. In other words, if we could divide the number of lives that could be saved because of these new technological platforms by the number of engineers and scientists that made them possible, you may find that the number of lives saved per engineer (or scientist) could surpass by far the number of lives that could be saved by any single medical doctor. So, if your goal is to save lives, the following chapters may convince you that sciences and engineering could both be very good options. In this respect, I sincerely hope that this book will be a source of inspiration.

To maintain the focus of the book, the following chapters put an emphasis only on the main technologies and principles used in this particular new interventional facility while comparing them to various scenes of the movie, just to make it more interesting and stimulating. Although the chapters do not expand much beyond that, the readers should be aware that there are many more technologies out there that are very amazing. The readers should also keep in mind that the technologies being presented are in constant evolution, and it is most likely that they will expand beyond what is briefly described in this book. You will also notice that I do not mention any names. This is because I do not want to forget anyone who collaborated directly or indirectly to the achievement of this new interventional facility. Without them, I would not have been able to write such a book. Some of them who played a more scientific role are listed as coauthors in one or more of my scientific papers.

The following chapters are really about this new interventional facility dedicated to cancer therapy that mimics the scenario of the movie. Unlike other approaches that have been investigated and that rely on systemic delivery (drifting out of control in the bloodstream), all technologies presented here have been developed with one goal in mind: to navigate therapeutic agents from the injection site (point A) to the tumor site (point B), a scenario indeed very similar to the one presented in the movie *Fantastic Voyage*. But if just going from point A to point B mixed with a bit of imagination could help cure millions of cancers per year, as Einstein quoted (see first quote at the beginning of this book), think about what more imagination could do. In this respect, the following chapters may help you realize that Einstein may have been on the right track.

1

The Location

B EFORE INTRODUCING THE more technological aspects later in this book, I would like to start by presenting the physical location and the actual building where the various platforms and systems will be installed. I said "will be" since I am presently writing this book a few months before the final installation of the platforms to allow sufficient time for the book to be released at the inauguration day (August 24, 2016) of this new infrastructure. The complete infrastructure will be located in the interventional facility, which is part of the NanoRobotics Laboratory at Polytechnique Montréal. Not to worry too much since we have several components of the infrastructure already installed and the design and specifications of the remaining equipment have already been determined in order for them to be installed on time, i.c., exactly 50 years after the release of the movie *Fantastic Voyage*.

I believe that you may have a better overall perspective if you know better the surrounding environment where the new fantastic voyages will take place. Furthermore, while writing this chapter, I found the comparison with the movie quite interesting with many quasi-identical features. I let you judge by yourself.

A Microscopic Submarine in My Blood: Science Based on Fantastic Voyage
Sylvain Martel
Copyright © 2016 Pan Stanford Publishing Pte. Ltd.
ISBN 978-981-4745-78-9 (Hardcover), 978-981-4745-79-6 (eBook)
www.panstanford.com

1.1   Landing at Los Angeles Airport (Scene 00:00:48–00:01:05)

This is the first scene of the movie that I selected. In this particular scene, you see a TWA aircraft landing at Los Angeles (LA) International Airport (LAX). Since the modern version of the facility is being installed in Montréal, Canada, an aircraft (from another company since TWA does not operate anymore) bringing people scheduled to come visiting us, would most likely land at Pierre-Elliot Trudeau International Airport (YUL). I indicated in Fig. 1.1 on a Google map, where the initial vision of Hollywood ends up 50 years later.

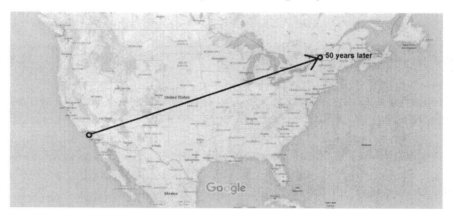

Figure 1.1 Starting in Los Angeles, where the fictional adventure *Fantastic Voyage* took place, it ends up 50 years later in Montréal, where the new real adventures à la *Fantastic Voyage* become a reality.

1.2   Transiting from the Airport to the Facility by Car (Scene 00:03:05–00:04:03)

In this particular scene, the car is being attacked while transiting from the airport to a secret facility. If they would have landed in Montréal, it would have taken roughly half an hour by car (around 20 kilometers) to make the trip from the airport to the new interventional facility of the NanoRobotics Lab at Polytechnique Montréal, which is located on the Campus of the University of Montréal (Fig. 1.2). This could have represented plenty of time for an ambuscade similar to the one in the movie, although this is unlikely in Montréal under normal conditions, knowing its reputation as a very safe city.

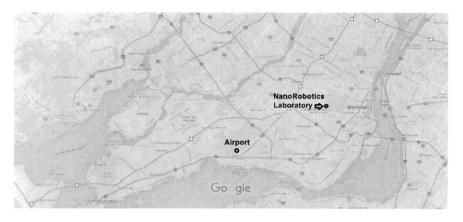

Figure 1.2 The locations of the airport and the NanoRobotics Lab being shown on a Google map of Montréal.

1.3 Secret Location of the Facility (Scene 00:05:48–00:06:58)

Although in the movie, the exact location of the facility is kept secret, the location of the new real infrastructure is much easier to find (Fig. 1.3). But, as in the movie, a car would have to go underground where there are parking spaces available. But no elevator is available

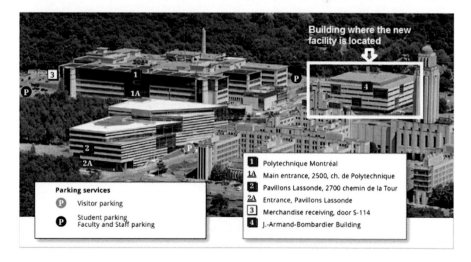

Figure 1.3 Photograph of the university campus showing the building (in a white square) where the next fantastic voyages will occur (adapted from http://www.polymtl.ca/rensgen/en/coordonnees/campus.php).

to bring your car down. Instead, you must drive downhill to access the parking spaces.

The Bombardier building depicted in Fig. 1.4 is home of the interventional facility of the NanoRobotics Lab. As you can see, it is easy to find once you are in the university campus. But people in general do not expect to see in this research building being part of an engineering school, an infrastructure containing a real medical interventional room containing exactly the same magnetic resonance imaging (MRI) scanner and other clinical systems typically found in hospitals. This alone could have explained why an elevator to bring a car in a secret underground facility similar to the one seen in the movie was not necessary!

Figure 1.4 Photograph of the Bombardier research building where the interventional facility of the NanoRobotics Lab is located. The access to the car entrance to the underground level is on the left side of the photograph, while the interventional facility is located on the ground (main) floor on the right side of the photograph and at the far end of the building (in the back, not visible here).

1.4 Entering the Underground Level (Scene 00:07:00–00:07:15)

It may look a bit weird to put a photograph of the doors of a garage (Fig. 1.5), but it is necessary if the goal of this book is to compare and to show how things have evolved since the making of the movie. In

Figure 1.5 Photograph of the doors for the car entrances and exits to and from the underground level.

this scene, Grant exits the car and enters the underground floor after a large door opens. In our real version, the car enters the underground floor through a large door shown in Fig. 1.5, once opened.

As depicted in Fig. 1.5, there is a box before the entry door where you must use a magnetic identification card to activate the door. This replaces the Military Police (M.P.) meeting Grant towards the end of the scene. You may have also noticed that the sign on top of the large doors is written in French. Indeed, this is another difference with the movie where English has been replaced by French. For information, the word "stationnement" means "parking." Although the majority of people in Montréal speak French, a large proportion of them also speak English. In fact, both languages are regularly spoken.

1.5   Transiting through the Underground Floor and Going to the Upper Floor (Scene 00:07:22–00:07:57)

In this scene, Grant accompanied by the M.P. travels through the underground floor and proceeds to the next upper floor. Figure 1.6 depicts how the other side of these large doors looks like once you are at the underground level of the Bombardier research building.

Figure 1.6 Photograph of the large access doors to the underground floor seen from inside the building.

As Grant proceeds through the underground floor, pay attention to how it looks, especially the vertical columns everywhere. Now pay attention at how it looks at present (Fig. 1.7) in the underground floor of our modern research building. As you can see, not much has

Figure 1.7 Photograph of the underground section right underneath the interventional facility where the next fantastic voyages will take place.

changed in the last 50 years for this particular aspect of the movie. The part of the underground that you see in Fig. 1.7 is exactly underneath the new interventional facility of the NanoRobotics Lab. Similar to the movie, this part of the facility does not look very fancy. During the last part of the scene, Grant and the M.P. use a ramp to proceed to the next upper floor. In the real version, you can use the stairs although most people use the elevator.

1.6 Mechanical Stairs and Adjacent Facilities (Scene 00:07:45–00:07:52)

As Grant transits from the ground floor to the main floor, you can see in the background, people using the mechanical stairs (escalators). It is interesting that mechanical stairs are widely used at Polytechnique Montréal as well. Figure 1.8 depicts on example of that. Here no people are shown to remain anonymous. During rush hours, these stairs are packed mostly with students and staff. This picture was taken in another building of Polytechnique Montréal known as the Lassonde building (building no. 2 in Fig. 1.3). The building is located on the same campus and next to the Bombardier research building.

Indeed, another difference with the movie is that the facilities are not in a single building but split between two main buildings. If you

Figure 1.8 Example of mechanical stairs used to transit throughout the building.

Figure 1.9 The photograph shows inside the section of the NanoRobotics Laboratory where most of the research and development are being done before being tested in the interventional facility. The computers are used mainly for emails, searches, designs, programming, simulations and modeling, writing scientific papers, etc.

look directly to the right from the middle of the mechanical stairs in Fig. 1.8, you will see three posters on three large windows. Behind these three large windows is located another part of the NanoRobotics Laboratory where the main research is being done before being tested in the interventional facility located at the Bombardier building (Fig. 1.4). Figure 1.9 shows this section of the NanoRobotics Laboratory, but without the persons in front of their respective computer. It is in this section of the NanoRobotics Lab that most of the new concepts are initiated. Other facilities (not shown here) throughout the campus but not being part of the NanoRobotics Lab are also used for various tasks such as machining, electronic microscopy, characterization and measurements, chemistry, etc.

At the end of the room, there is another small meeting room where we do brainstorming. We often use a white board to express our ideas. This board is continually updated with new concepts and ideas. Figure 1.10 shows an example of what the board looks like. I took this photograph while I was writing this page, so it is not biased, and very often, the board looks even more chaotic and busy than this one. As

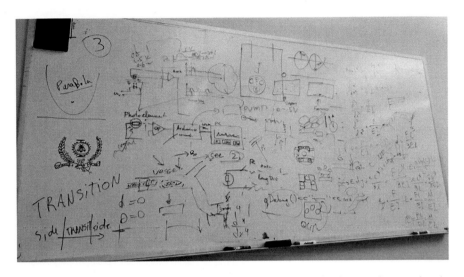

Figure 1.10 Photograph depicting a white board after a brainstorming session in the meeting room of the NanoRobotics Lab.

a note, my office is just next to this section of the lab, and inside, it looks even more chaotic. I just show in Fig. 1.11 the small board in my office. I do not show the rest because I do not want to scare you!

Figure 1.11 Photograph depicting the white board in my office where initial ideas often end up as real implementations.

1.7   Security Desk at the Entrance (Scene 00:07:52–00:08:00)

In this scene, you see a security agent at her desk. This is the same when entering the main (ground) floor of the Bombardier building, just above the parking on the underground floor. Here, the equivalent security desk at the entrance of the main floor of the building is shown in Fig. 1.12. As in the movie, no ID is typically being checked here (although the agent maintains a surveillance), but this will change at the next security level described in the next section.

Figure 1.12 Photograph of the more modern security desk at the entrance on the main floor (at same level as the interventional facility of the NanoRobotics Lab).

1.8   Identification Check (Scene 00:08:00–00:08:17)

In this scene, Grant gives his ID card to the security agent before he introduces it in some kind of card reader. The technology has evolved since then. Indeed, as depicted in Fig. 1.13, the security agent has been eliminated and replaced by the reader located on the left of the main entrance of the interventional facility. You do not need to introduce the card anymore. As it is also done in many facilities, you just need to put the card in close proximity of the magnetic reader. When this commonly used electronic security system detects that the owner is

Figure 1.13 Photograph of the main entrance to the interventional facility of the NanoRobotics Laboratory. Controlling the accesses to the facility is done by the electromagnetic reader depicted as a small black box on the left of the image and next to the door.

granted access to the facility, the door unlocks automatically instead of the gate in the movie that lifts to allow entrance. But the goal shared by both versions remain ultimately the same: controlling access to the facility.

1.9 Transit towards the Facility (Scene 00:08:18–00:08:38)

Figure 1.14 shows what you would see in the real version when transiting from the front security desk (Fig. 1.12) to the main entrance of the interventional facility of the NanoRobotics Lab. As one can see, the building really looks more modern compared with what is being showed 50 years ago in the movie. The main entrance of the interventional facility is located at the far end of the corridor depicted in Fig. 1.14.

Figure 1.14 Photograph of a section of the Bombardier building leading to the main entrance of the interventional facility of the NanoRobotics Laboratory.

1.10   Signs and Security at the Entrance of the Facility (Scene 00:08:38)

In this scene, you can see several signs such as "Section B," "Unit 7," and "Medical Division." As you can see in Fig. 1.13, the sign for the main entrance of the medical interventional facility of the NanoRobotics Lab is much more discrete compared with the Hollywood version. The small and only sign depicted in Fig. 1.13 indicates just basic information such as the local's number and security information related to potential exposure hazards. In this respect, I think that Hollywood did a better job!

2

The Interventional Facility

A FTER I BRIEFLY showed the location and the surrounding environment, we are now proceeding to the interventional facility itself. In this chapter, you will also see some similarities with the movie and some differences as well that I will explain along the way. The first scene of this chapter begins when Grant enters part of the facility where the patient is being prepared. This is also where Grant meets General Carter for the first time.

2.1 The Preparation Room (Scene 00:08:38–00:09:17)

In this scene we can see the patient, Dr. Jan Benes, being prepared for the medical intervention consisting of injecting a miniature submarine in his carotid artery. The images of the scene can be compared to the one in Fig. 2.1, depicting the preparation of a pig used here instead of a human where we are testing a new medical protocol for injecting miniature "submarines" in an artery prior to navigate them towards the targeted physiological location. As you can see, it is somewhat similar, although in Fig. 2.1 you can see the modernization of the medical systems over the years.

A Microscopic Submarine in My Blood: Science Based on Fantastic Voyage
Sylvain Martel
Copyright © 2016 Pan Stanford Publishing Pte. Ltd.
ISBN 978-981-4745-78-9 (Hardcover), 978-981-4745-79-6 (eBook)
www.panstanford.com

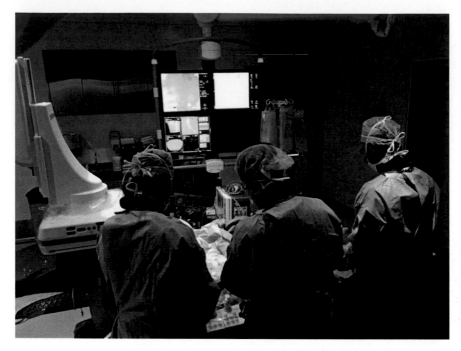

Figure 2.1 A team prepares the patient (a pig) prior to the injection of miniature "submarines" in an artery that will be followed by the navigation phase in the bloodstream.

Indeed, we are using pigs for preclinical studies since these animals have anatomical features that resemble the ones found in humans. They are presently used since the first mission of the new interventional facility is translational research. In other words, its first mission is to translate, with the participation of engineers, scientists, and medical doctors, new medical interventional platforms and related technologies to the clinics. As such, the intervention must be fully tested and validated first on animals for its efficacy and safety prior to being used on humans. As you can observe in Fig. 2.1, the room and the medical systems being used for the preparation phase are similar to the ones found in modern clinical interventional rooms. This will not be entirely the case during the interventions, since new platforms never seen in a clinical settings so far are being integrated in this new interventional facility.

2.2 First Look at the Interventional Facility (Scene 00:09:17–00:10:50)

In this scene, when entering the monitoring room, other members working in other rooms are contacted through TV monitors. In this scene, we realize that several rooms constitute the whole interventional facility.

The fact that the facility in the movie is constituted of several rooms mirrors the same approach used for the implementation of the NanoRobotics facility. This is shown in Fig. 2.2, depicting a schematic of the interventional facility of the NanoRobotics Lab. Other sections are physically located at various locations as in the movie. One example of that is the main R&D section of the NanoRobotics Lab (Fig. 1.9).

As depicted in Fig. 2.2, the interventional facility consists of four main rooms (other rooms are used for storage), namely, the preparation room, the control room, the interventional room, and the electronic and mechanical room. As in the scene of the movie showing the patient being prepared, the preparation room is where the patient (or animal) is being prepared before entering the interventional room. The interventional room is where the actual navigation in the bloodstream takes place. In this room, there are platforms that you will find in a hospital such as the clinical magnetic resonance imaging (MRI) scanner and the x-ray platform, to name but two examples, but also new interventional platforms that you have probably never heard of and that are nowhere else to be found. They have been developed at the NanoRobotics Lab especially to make the whole scenario of the movie possible which was a must in order to maximize the treatment efficacy achievable with this new facility. The role of these platforms will be explained later in this book.

Adjacent to the interventional room is the electronic and mechanical room. It contains the electronic cabinets for the various platforms in the interventional room. Some cabinets are used for imaging inside the body and to track the "submarines," while other electronic cabinets are dedicated to other various tasks such as the ones related to the navigation of these "submarines," communicate information to the "crew members," and even to open barriers (blood–brain barrier

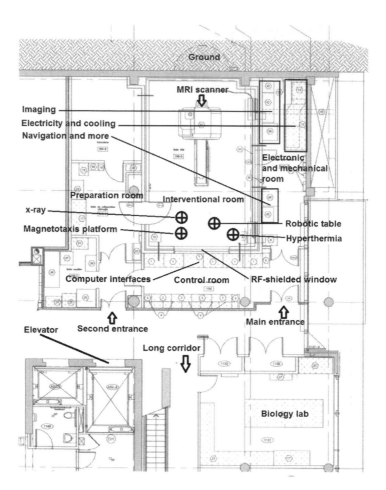

Figure 2.2 Plan of the NanoRobotics Lab interventional facility with the various rooms.

[BBB]) to access the interior of the brain allowing us to reach the same physiological region envisioned in the movie. The room also contains other systems such as the electrical panels to distribute the high power required by these platforms and the plumbing related to cooling, which is required to maintain the temperature of these high-power units within an acceptable temperature range. Finally, the control room is where the staff responsible for the coordination of the mission are located. They operate in front of special computer displays while keeping a visual on the interventional room through a large meshed window serving also as an RF shield. If you face the arrow indicating the long corridor in Fig. 2.2, with the entrance of the biology lab to your right, you will see what is shown in Fig. 1.14.

You may have also noticed another room in the far right corner being identified as the biological lab. This lab is also part of the NanoRobotics Lab facility but is not part of the interventional facility. You will understand later the reason for the need of a microbiology lab.

2.3 Name of the Facility (Scene 00:10:50–00:11:27)

At the beginning of this scene, Grant looks at the large sign on the wall showing the characters CMDF before attempting to guess its meaning in a humoristic way. Later, while proceeding to the next destination, General Carter spells out what those letters mean and provides an explanation behind the meaning. If we would post our logo as in the movie, we would most likely have the one depicted in Fig. 2.3.

The logo shows atoms being connected together by a frame forming the letter "N" for Nanorobotics. The two semicircles indicate that the atomic structure can move, suggesting a robot made of connected atoms. The scale represented by "nm" for nanometer suggests the motion control of a nanometer-scale structure, like the ones used to implement the technology for navigating in the bloodstream.

In the prologue, I briefly explained the name NanoRobotics, linking the word *robotics* with *nano*, the latter suggesting very small

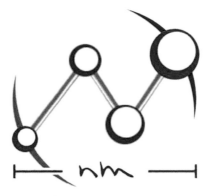

Figure 2.3 Present logo of the NanoRobotics Laboratory showing four circles representing atoms being linked together and forming a nanometer-sized robotic structure in motion. The structure forms the character "N" for Nanorobotics.

objects. This is indeed true, but it goes far beyond just the overall size of an object. More precisely, it refers to nanotechnology that exploits the physic phenomena that only occur at such a scale.

Nanotechnology refers to the application of extremely small things and can be used across all fields of science such as chemistry, biology, physics, material sciences, and engineering. While attempting to make the movie *Fantastic Voyage* a reality, we needed to push engineering beyond what was practically possible using more traditional engineering approaches.

Indeed, engineers develop new technologies or systems that rely on principles or laws of physics. Now, think of engineers trying to solve a problem or to create something inspired from a science fiction movie by using laws of physics contained in one book. Let's call it the "traditional physics book." After years of efforts, the engineers then realize that the range of possibilities provided by this traditional physics book is limited and constrains what they can actually realize. Engineers were then facing a technological dead-end until nanotechnology came along.

Activities in the field of nanotechnology were boosting at the time when the NanoRobotics Lab was founded 15 years ago, when new discoveries were revealed in scientific journals. When I tried to develop the first principles aimed at attempting to make the idea behind the Hollywood movie a reality, it was exactly like if I found another book that would reveal more laws of physics. Keeping the first book at hand, I would use many of the new laws of physics inside this new book to expand the range of possibilities available to me in order to create what was so far believed to be impossible. It was like magic made possible by nanotechnology.

Indeed, people do not realize how powerful and magical, in a sense, nanotechnology can be. It is like discovering a new world of possibilities. This is because objects have different properties and behave differently at the nanometer scale.

To put these dimensions in perspective, a human hair is approximately 100 micrometers thick. You divided it by one hundred and you get one micrometer (μm). One micrometer is approximately the size of bacteria, which are even smaller than a red blood cell that measures a few micrometers across. Now, divide the size of a

1 micrometer bacterium by 1000 and you get 1 nanometer (nm). One nanometer corresponds to only ten hydrogen atoms (the smallest atom in the universe that we know so far) aligned side by side. It is rather small. To help further put that in perspective, a sheet of newspaper is approximately the same thickness as a human hair, divide that by 100,000 and you get 1 nanometer.

So, the facility is called the NanoRobotics Lab because it links the principles of robotics with nanotechnology to implement things like a small "submarine with its crew" capable of navigating in the bloodstream. We could indeed do that with nanotechnology instead of miniaturizing a larger submarine and its crew members, as the movie suggests, since this approach was and still is well beyond technological feasibility. You will see later in this book how nanotechnology has made it possible.

2.4   Electronic Cabinets (Scene 00:11:48–00:12:00)

In this scene, as Grant transits along the corridor with General Carter before going up through mechanical stairs similar again to the ones shown in Fig. 1.8, you can see several electronic cabinets defiling in the background. In the real facility, there are indeed electronic cabinets but they are installed in the electronic and mechanical room depicted in Fig. 2.2. There are many of those as in the scene in the movie and the front panel for each of them has a different look, depending upon the specific purposes they serve. The photograph on the left of Fig. 2.4 shows the electronic and mechanical room as it appears at the time of the writing. It will look much more crowded as more electronic cabinets are planned to be installed in the future.

An example of an electronic cabinet is depicted in the middle photograph of Fig. 2.4. This cabinet is used for imaging and tracking inside the body and to gather the information required for navigation in the bloodstream.

Another example is depicted in the right photograph of Fig. 2.4. It shows the front of an electronic cabinet involved in the navigation control of the "submarines" in the bloodstream. Several other cabinets exist but are not shown.

Figure 2.4 Photographs showing part of the electronic and mechanical room (left); an example of an electronic cabinet that is dedicated to gather physiological images such as the blood vessels to be navigated, and for tracking the "submarines" in the body (middle); and another example used for particular navigation tasks.

The present placement of these electronic cabinets is shown in Fig. 2.2. These photographs give you a basic idea of the supporting hardware that is required to make the navigation of "crewed submarines" in the bloodstream a reality.

3

The Team at the Planning Briefing

N OW THAT YOU have a better idea of where the action will occur on both sides, the more scientific aspect begins in the following sections. In this chapter, we go through the scenes of the movie describing the plan of the mission. During the briefing, we will have the chance to meet the team members. All this will be excuses to write additional lines of text that will compare the scenes of the movie to the real modern version while adding comments on the importance of having good team members to successfully accomplish this type of complex missions.

3.1 Maximum Time to Accomplish the Mission (Scene 00:11:35–00:11:38)

In this very short scene, General Carter mentioned to Grant that they can only maintain objects miniaturized for 60 minutes. This constrains the maximum time allowed to complete the mission to a maximum of one hour. This is particularly important for the crew members of the submarine since they must exit the body of the patient, Dr. Benes, before they begin to resume to their original size.

A Microscopic Submarine in My Blood: Science Based on Fantastic Voyage
Sylvain Martel
Copyright © 2016 Pan Stanford Publishing Pte. Ltd.
ISBN 978-981-4745-78-9 (Hardcover), 978-981-4745-79-6 (eBook)
www.panstanford.com

It is interesting fact that our "miniature crew members" also have up to approximately 60 minutes to accomplish their mission. Unlike the crew members in the movie who must come back alive, our "crew members" go on a suicide mission, but they do not know about it. I agree that not telling them sounds really bad. But wait until later in this book to find out how many crew members go on such a suicide mission. You may be more outraged when you find out how many. What I can tell you now is that it is way more than in the movie.

Without any attempt to extend their life, once injected in the body, our so-called crew members will die after approximately 30 to 40 minutes. But we can possibly expand their lifespan to 60 minutes using special techniques, but, as in the movie, this is probably about the maximum time that they have to complete their mission. As in the movie where they try to get the secret to extend this time to allow them to apply the technology to other types of missions, we also do not know at the present time how to extend the time beyond the one hour limit. Although we found over the years and after several experiments that such time is sufficient to accomplish missions similar to the one in the movie, being able to extend their life in the body would allow us to consider, as the movie suggests, other types of missions as well.

3.2   Goal of the Mission (Scene 00:12:06–00:12:14)

This very short scene shows General Carter divulgating to Grant the ultimate goal of the mission. At his surprise, Grant learns that they will shrink a submarine with its crew members and inject them in an artery to reach a clot that formed inside the brain of Dr. Benes.

Although this may sound crazy, the short description provided by General Carter really fits the approach being used by our team, but with just some minor modifications. For instance, we typically use more than one submarine and we do not shrink any crew members, since we use already miniature crew members who will still obey our commands before and after they have been injected in the artery. Besides that, the plan described by General Carter amazingly matches how we do it half a century later.

3.3 A Woman in the Team (Scene 00:12:40–00:13:10)

This scene really shows how the role and the perception of women in sciences and engineering have evolved. In the movie, only one woman (Cora) is the topic of argumentations as if she should or should not be part of the team. It is also mentioned that this is not the place for a woman.

To be honest, I did not have a single woman in my team during the first few years of this adventure. It is really not because I am sexist, but because not a single woman asked me to join the team. But soon after we announced to the world the first successful navigation of an untethered object in an artery, insisting that such a demonstration would lead to a new approach to cure cancers and therefore potentially save human lives in the future, then surprisingly, I began to receive tons of requests from women wishing to join my team. What I learned over those years is that girls can be as good as boys and even better sometimes at anything, including engineering and sciences. This is especially true if this is for a humanitarian cause, such as saving the life of humans diagnosed with cancer. Since then, approximately 50% of our team members are women, which is not bad for an engineering school!

The other difference with the movie is that I also have people from different origins and appearances. Over the years, I had people with darker skin, Asians, and people from various countries and continents. I always said that I do not care how they look. If they had blue skin and antenna on their head and came from another planet, it would be fine with me if they would be motivated, creative, reliable, willing to work hard, and could be a good team member.

3.4 Team Members (Scene 00:13:10–00:13:44)

In this scene, General Carter introduces the team members. As you can see in the movie, each member has a different background and expertise. In reality, to conduct such a complex mission, we need a much wider interdisciplinary team. Yes, we have medical doctors in the team as well with the specialization corresponding to the type of cancer, the location in the body, and the types of interventions that will be performed. For instance, if we target the liver, or a colorectal tumor, or a tumor in the brain, etc., we will have a different medical specialist in the team. This could be an interventional radiologist for the liver to an oncologist specialized in brain cancer. But many other scientists are presently part of the team as well. We already have various biochemists with different specializations, microbiologists as well, etc., working with engineers from various specialties, mathematicians, and computer programmers. The list goes on and on.

You have to be multidisciplinary yourself if you want to survive and lead in this type of environment. For instance, I did a PhD in electrical engineering following four years of studies to obtain a bachelor's degree, an additional two years in computer sciences (CS), and two years for a master's degree. Then I spent 10 years in the Department of Mechanical Engineering at MIT, and now I am in the Department of Computer and Software Engineering at Polytechnique Montréal. Despite all this time in universities learning new topics, these days, I am involved in cancer therapy talking about topics ranging from drug formulations to specialized medical interventional procedures that have nothing related to what I learned at school. This is to say that it does not matter how much you learned and how long you stayed at school, to survive and to perform in such an environment, you need to learn every day while keeping a very open mind, and not just stick on the specialty that you studied while at school.

Indeed, this will allow you to talk the same language as other specialists while remaining expert in your field, which will most likely prove to be also useful. Indeed, I found that communicating effectively with team members from other complementary disciplines proved to be one of the hardest things to achieve. If you do not learn the vocabulary associated with each specialization, and you do not know at a minimum the fundamentals associated with each specialization, you will not seek answers to the right questions and you will miss opportunities. This is true even if you have a very creative mind, which by the way is another quality that we are looking for in students who want to be part of the team.

When someone asks me how I see my role in such a multidisciplinary team, I respond by comparing myself to a conductor of an orchestra. Indeed, the conductor cannot be the best at every type of instruments (specializations) in the orchestra (the team), but he knows music (sciences and technologies) very well and what each musical instrument can do. He also knows which instruments should be included and how and when to use them so that the music sounds like a harmony. Without the conductor to coordinate all these musicians, the result would sound like a cacophony. This is pretty much the same here where all the right instruments must be assembled to form one orchestra and where each team member (musician) should know how to play well their respective instrument. But they should also know how to play in a group. I should add that if the music was composed by the conductor himself, it is even better and often a must in a research and development facility.

Going back to the movie, towards the end of the scene, General Carter also introduces a naval officer who will be part of the team and, as you will see later in the movie, will be responsible for driving the submarine in the bloodstream. Well, it may be another coincidence with the movie, but I was a naval officer myself. During my 29 years in the Naval Reserve while studying or working at the university, I have fulfilled many positions in the navy including but not limited to diver and navigator and, towards the end of my career, as a commanding officer for eight years. But I was in command of a military ship, not a submarine. I can say (with a wink) that not only the number of crew members (35 to 45) on board my ship surpassed by far the five crew members in the miniaturized submarine in the movie, but navigating in the Pacific and especially in the North Atlantic Ocean is most likely more prone to sea sickness than navigating in the bloodstream, even within the turbulent flows when the heart is pumping blood at full rate, believe me on that!

3.5 Beginning of the Briefing (Scene 00:13:52–00:14:23)

As in this scene, we also have several briefings with the team members especially before an intervention. Sometimes, we use Skype on a computer display instead of the TV monitors that were shown in the scene 00:09:17–00:10:50 (Section 2.2), or simply by phone through a speaker so that everyone in the meeting room can hear. Indeed, among all the team members, many are from other institutions such as universities and hospitals, and getting everyone to a single physical location is often very difficult considering the very busy schedules of our team members.

The other difference with the movie is the use of transparencies and the related projector that, as you know, are now obsolete. As others do, we typically project a PowerPoint presentation on a screen (or on one of the walls of the meeting room where there is no screen) that runs on a computer or a laptop being connected to an LCD projector.

Another funny moment (00:14:10) of the movie is two guys smoking big cigars in the meeting room. The only resemblance with us is that I also have two guys in the team who smoke, but not in the meeting room, and even not inside the whole building, as smoking is not permitted anymore except outside the building.

The meeting gets more interesting when it is mentioned that the submarine will be injected in the carotid artery (scene at 00:14:22). I found it very interesting because our first successful in vivo navigation in the bloodstream was done also with only one "submarine" and, also by pure coincidence, in the carotid artery. I say pure coincidence since it is after the publication of the scientific paper showing our results confirming the first automatic navigation of an untethered object (a 1.5 mm chrome-steel bead representing our initial and primitive version of the submarine) in the blood vessel of a living animal that the movie *Fantastic Voyage* was referred to for the first time in this context in a press release published in 2007.

In the scene of the movie (00:14:15), you see a projection of the arterial network that is used to explain by which route the submarine will travel to reach the brain. Although general images of the vasculature are often used for discussion, typically we gather the real images of the artery of the patient (here being a pig) that will be transited during the intervention. The reason is that there are small differences between patients. So we need to do that to achieve a better planning phase that will enable a more accurate and reliable navigation phase. This could be seen as a kind of personalized medicine. Figure 3.1 depicts an example of an image gathered for planning purpose. Typically, images of the larger blood vessels are taken by injecting contrast agents that can be detected by x-ray. This is known as angiography.

From the images obtained such as the one depicted in Fig. 3.1, several transformations can be done. Figure 3.2 shows one example

Figure 3.1 Example of an image of blood vessels gathered for planning purpose.

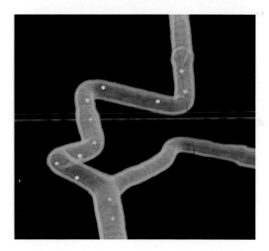

Figure 3.2 One example among many possible examples on how the blood vessels can be represented on a computer display where the planned trajectory of the "submarine" is being displayed as a series of dots being referred to as waypoints.

where commands indicating the route to be followed are represented by successive dots also being referred to as waypoints along the planned navigation path. Image processing can be used for many purposes such as enhancing the image representation and to remove all blood vessels that are not intended to be transited.

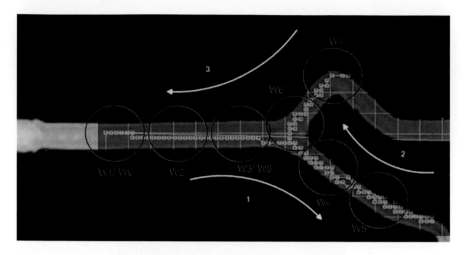

Figure 3.3 Another example among others where waypoints are superimposed along the planned trajectory in an artery to allow automatic navigation control to be executed by a computer from the position of the miniature "submarines" being tracked by a clinical MRI scanner. (From *Magnetic Resonance in Medicine* 59:1287–1297, copyright © 2008, with permission from John Wiley and Sons.)

More advanced representations are also possible. Figure 3.3 depicts one example that we investigated and where waypoints represented by circles that can be superimposed on blood vessels intended to be transited before reaching a predetermined physiological target. The information plotted in this image is used by a control computer to navigate automatically one or more tiny "submarines" along the successive waypoints from tracking information gathered by a clinical MRI scanner. The computer's decision in this example is the same within each circle and is typically different for each circle. The use of circular waypoints of specific diameter accounts for small delays and motion artefacts that can occur during the navigation phase. The small squares along the planned trajectories are the tracked signals of the miniature "submarine" acquired during the navigation phase.

Figure 3.4 shows the waypoints plotted along the carotid artery that were used by the computer to automatically navigate for the first time an untethered object in the bloodstream of a living animal at an average velocity of 10 centimeters per second. This was done one late evening in November 2006. Right after completing this world première, the boss (i.e., me) was happy to pay the rounds.

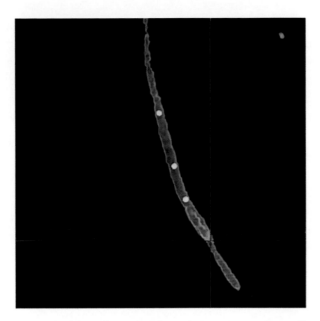

Figure 3.4 Image showing the waypoints that were used to allow a control computer to navigate for the first time an untethered object along the carotid artery of a living animal.

3.6   Blood Flow Rate and Its Potential Impact on the Miniature Submarine (Scene 00:14:23–00:14:54)

This scene puts in perspective a very important issue when it is mentioned that at such a scale the submarine once miniaturized will cruise extremely fast because of the blood flow. In fact, on the basis of our experience, it is way too fast, especially in arteries, including the carotid artery. In the movie, it is proposed to slow down the heart rate of the patient in order to decrease the blood flow rate. In our case, we presently decrease the blood flow rate using a balloon catheter. Once inserted in the artery, the balloon catheter is inflated, which causes a reduction of the blood flow sufficient to enable proper navigation.

It is also mentioned that they will use hyperthermia to slow down the heartbeat rate. I am not sure if I heard this right, maybe not. Since hyperthermia is related to an elevation of the temperature level, I suspect that the movie meant hypothermia instead, since they are talking about lowering the temperature level of the patient. It's interesting here that we were discussing of potentially testing hypothermia to lower the temperature level of the physiological regions where we would operate, not for extending the life of the patient but for extending the life of our microscopic "crew members" to roughly 60 minutes, as in the movie. Nonetheless, hyperthermia is also used for specific missions, as well as to miniaturize some types of submarines that are designed to navigate in the arteries, and to open the blood–brain barriers to allow access to deeper regions of the brain, as in the movie. I will explain more on this later in this book. At the end of the scene, it is mentioned that they do not plan to go through the heart. This is also in accordance with our practices.

3.7 Recovery Route (Scene 00:14:54–00:15:26)

In this scene, it is mentioned not only which vascular route will be used to get to the brain, but also which route will be used to get to the recovery location. When we did navigation for the first time, it was also as I specified earlier in the carotid artery. In this respect, the navigation path was similar to what has been proposed in the movie. Indeed, during this particular experiment we had a recovery point, but it was still in the carotid artery and not in another blood vessel, as the movie proposes. As a short note, we still have this miniature "submarine" that was recovered during this experiment demonstrating for the first time that navigation in the bloodstream was possible. After our first proof of concept, the "submarines" became much smaller and recovery was not an option anymore.

3.8 Tracking the Position of the Submarine (Scene 00:15:30–00:15:46)

In this scene, Grant asked how they will know the position of the submarine. It is mentioned that the submarine will be nuclear powered and that the radioactivity will be used to track the position of the submarine.

Yes indeed, the position of a miniature submarine with radioactive tracers could be detected with a technique known as positron emission tomography (PET). That could also be an option for us. But because our submarines are not nuclear powered, we generally use MRI instead, although in some cases, the use of PET could also be an option.

3.9   Danger of Attack (Scene 00:15:56–00:16:18)

In this short scene, it is mentioned that the submarine and its crew are in danger of attack by the natural defense system of the body. This is indeed an issue for our submarines, the crew members, and any objects that we plan to inject in the bloodstream. We are doing many tests to make sure that all these objects are biocompatible. We also try to design submarines with components that we already know that they are biocompatible. Furthermore, we are travelling towards our targeted physiological locations as quickly as possible before the occurrence of a potential attack that could prevent the success of the mission.

3.10   Sense of Humor (Scene 00:16:20–00:16:30)

In this short scene, Grant makes a final joke at the end of the meeting. Indeed, we often make jokes as well. Indeed, when you face problems and challenges all the time, having a good sense of humor really helps keep the moral up. As such, having a good sense of humor is a must if you want to survive and be part of our team. At the end of the meeting, they proceed for sterilization, which, of course, is a common practice when you plan a surgical procedure, although we do not have a sterilization room similar to the one depicted in the movie.

4

The Submarine

A FTER THE MEETING and seeing more the interventional facility, including the control room, and the patient being prepared while its vital signs are verified, which by the way is also done in our facility, the submarine finally appeared in front of the crew members when they exited the sterilization room. This chapter is entirely dedicated to the submarine.

4.1 External View of the Submarine (Scene 00:18:31–00:18:35)

In this scene, we can see the submarine for the first time. Although our submarines can take various forms, Figure 4.1 depicts one example that was used a few years back. It was navigated in the hepatic artery of living rabbits in order to reach deep regions in the liver. The same type could also be used in other blood vessels including the carotid artery of a human as in the movie.

4.1.1 *Crewless Submarines*

The version depicted in Figure 4.1 was synthesized following the first version that was used to demonstrate the first navigation of an

A Microscopic Submarine in My Blood: Science Based on Fantastic Voyage
Sylvain Martel
Copyright © 2016 Pan Stanford Publishing Pte. Ltd.
ISBN 978-981-4745-78-9 (Hardcover), 978-981-4745-79-6 (eBook)
www.panstanford.com

Figure 4.1 Microscopy image of one type of "submarine" that was used to navigate in the hepatic artery towards deep regions in the liver.

untethered object in the carotid artery. This first version was also round in shape but was much larger with an overall size equivalent to the tip of a ball pen. It was metallic as I am guessing, the body of submarine in the movie. Although I do not know of which metal the submarine in the movie was probably made, our first version was made of chrome steel. The second version depicted in Fig. 4.1 was made of various materials as the submarine in the movie, with a polymer (plastic) being used as the main material. This might also be the case for the submarine in the movie.

The second version depicted in Fig. 4.1 was indeed much smaller with a diameter of approximately 50 micrometers (μm), about half a human hair thickness as shown in Fig. 4.2. Such a size allowed navigation in narrower blood vessels since the first larger version was restricted to large arteries only.

The overall diameter could even be reduced further. We also synthesized versions as small as 10 micrometers in diameter, but the propelling force produced to navigate effectively in the bloodstream was not as good as the larger versions. We estimated that the optimal diameter of such submarines intended for specific interventions in humans would be between 150 and 200 micrometers but not much larger since it would restrict the regions in the body that they could operate. Larger submarines can be designed to produce larger propelling forces than smaller ones. So our goal is to develop the largest possible submarines but sufficiently small to be able to reach

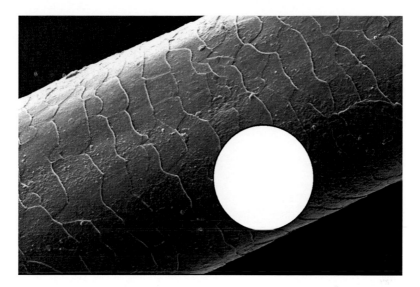

Figure 4.2 Overall dimension of the "submarine" being shown as a white circle on a human hair.

deep physiological regions accessible through narrower blood vessels. For crewed submarines, this would be sufficiently close to allow the microscopic crew members to swim to the region to be treated.

4.1.2   *Crewed Submarines*

The previous "submarines" did not have crew members onboard. They could be considered as miniature drones being controlled in the bloodstream from a far location (here being from the control room, Fig. 2.2). Figure 4.3 depicts a version of a submarine capable of carrying a crew. The miniature crew members inside the submarine (depicted as small dots inside the sphere) are still alive.

   Although the version depicted in Fig. 4.3 and capable of carrying crew members is still under development, it gives an idea of how the future versions of the crewed submarine could look like. The dimensions of this version are comparable to the crewless ones mentioned earlier. But the material used to implement the external wall of the submarine is different. Although there are many construction (synthesize) materials and different methods that can be

Figure 4.3 One version of the navigable submarine with the crew members onboard seen as small dots inside the round assembly.

used, Fig. 4.4 depicts one example among others that has been used in the past by our team.

In Fig. 4.4, the external wall of the so called submarine is built from a chain of molecules forming a hydrophobic bilayer made of a molecular assembly that is constituted of hydrophilic head towards the exterior and hydrophobic tails towards the interior of the submarine. Hydrophobic means that it repulses water while hydrophilic is the opposite, i.e., that it attracts or retains water molecules. As you can

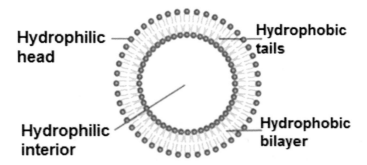

Figure 4.4 Construction of a miniature crewed submarine based on techniques used for the synthesis of liposomes.

see in Fig. 4.4, the external side of the submarine's wall is hydrophilic which is desirable considering that blood is a fluid. But the interior is hydrophilic as well meaning that unlike the submarine in the movie where the living space is filled with air, the living space inside the submarine for our crew members is completely flooded with an aqueous solution. This suggests that our crew members should be able to survive in a fluid environment, otherwise they would need to carry an oxygen tank like divers do, the latter not being a valuable solution for us. Other assemblies could be investigated including but not limited to micelles. This part of the building process is just pure biochemistry.

4.2   Entering the Submarine (Scene 00:18:35–00:19:18)

In this scene, we see the crew entering the submarine one by one through an entrance (hatch) located on top of the submarine. You may wonder how our crew members enter our miniature submarines, such as the one depicted in Fig. 4.3. The reality is that the crew members are injected into the miniature submarines during the fabrication (synthesis) of the submarines, before finalizing the implementation of the external walls of the submarines (Fig. 4.4). This is done through a setup containing a special microfluidic network. To keep the description as simple as possible, think of a miniature platform that has two capillary channels with widths around a human hair thickness that intercept in one location. One channel is dedicated to the formation (fabrication) of the submarines and the other one for injecting under pressure the crew members into the submarines before the crewed submarines exit the microfluidic platform.

4.3   Propulsion Power (Scene 00:19:46–00:20:45)

In this scene, we learn that the submarine is powered by nuclear energy and that such energy is provided by a microscopic nuclear particle. Furthermore, it is mentioned that they use a particle that is already small because nuclear particles cannot be miniaturized with their technology. It is assumed, watching the movie, that the power generated is used to provide the electricity onboard for lighting and to power the various instruments and the communication unit, but also, and not the least, to propel the submarine.

4.3.1 Microscopic Particles to Power the Submarines

What is interesting here is that the power of our submarines is also provided by microscopic particles. But since our submarines do not need onboard electricity, all this power is used for propulsion and steering (directional propulsion). Also in concordance with the movie, our particles cannot be miniaturized either from their original size. But one major difference is that our particles are not nuclear but magnetic.

For our first submarine navigated in the carotid artery, we used a single particle as in the movie. The round miniature particle was ferromagnetic and had a diameter of 1.5 mm (roughly the size of the tip of a ball pen, which is considered extremely large in medical nanorobotics). The magnetization of ferromagnetic material depends on that of the applied magnetizing field. In other words, increasing the magnetic field applied to the ferromagnetic particle is similar to increasing the power of the engine, whereas a permanent magnet has already a fixed power regime, called magnetization. But the question is, why use a ferromagnetic particle instead of a permanent magnet, the common magnet that most of us played with when we were young? The main reason is that the magnetization level of some ferromagnetic materials can exceed the maximum magnetization of a permanent magnet if the ferromagnetic particle can be exposed to a sufficiently high external magnetic field. It is like having a 12-cylinder engine instead of a 6-cylinder engine under the hood, and indeed we need all the power that we can get if we want to navigate in the bloodstream.

4.3.2 Superconducting Power

Such a high external field, however, cannot be generated by a simple magnet but with a superconducting magnet, which is much more complex to implement. This fact alone explains the need for the electronic and mechanical room depicted in Fig. 2.2, especially for the control electronics, the high power distribution, and the cooling infrastructure.

A superconducting magnet is an electromagnet made from coils of superconducting wires. As most of you already know, you can create a magnetic field by passing an electrical current in a wire. Increasing the electrical current will increase the magnetic field being generated. We refer to this field as electromagnetic field, which is a magnetic field generated by electricity and not by a permanent magnet. But by increasing the current to a very high level, the wires will begin to heat excessively until they break. To maximize the amount of electrical current while preventing any breakage, the wires are cooled at cryogenic temperatures during the operation. At such extremely low temperatures, the wires in the superconducting state can conduct much larger electric currents and hence create the intense magnetic fields required for our submarines. Liquid helium is typically used to cool most superconductive wires installed in thermally insulated reservoir called a cryostat (from *cryo* meaning cold and *stat* meaning stable) that maintains the low cryogenic temperature. An outer jacket surrounding the cryostat and filled with liquid nitrogen is typically installed to keep the liquid hydrogen from boiling away. This is the configuration that was initially installed in our facility (Fig. 2.2). Other recent superconducting systems are cooled instead using a two-stage mechanical refrigeration unit due to the increase cost and the limited availability of liquid helium. The latter configuration is the one that replaces the old one in the our facility.

4.3.3   *Ferromagnetic Propulsion Engines*

But we had a major problem with ferromagnetic particles, preventing them to be considered for clinical applications such as drug deliveries in cancer therapy. The following explanations may be a bit more difficult for some of you to understand, but keep reading. I will try to make the explanations as simple as possible.

Ferromagnetic particles have a remnant magnetization, also known as remanence or residual magnetization. This is the magnetization that remains in the ferromagnetic particles after the external magnetic field is removed. In other words, it is like if you want to turn off your engine but it does not stop and keeps running but at a lower regime. This is

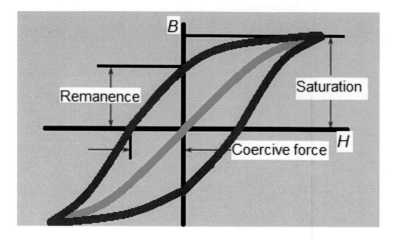

Figure 4.5 Typical magnetization curves of ferromagnetic particles in red and for superparamagnetic nanoparticles in green. As depicted in the graph, unlike permanent magnets or ferromagnetic particles, superparamagnetic nanoparticles have no remanence.

shown in Fig. 4.5, which depicts the typical magnetization curve of ferromagnetic materials plotted in red.

In the graph, the horizontal axis represents the magnitude of the external magnetic field (magnetic field strength) or magnetization field (typically represented by H) at the location of the magnetic nanoparticles (or if you prefer, at the location of the magnetic nanoparticle-powered submarine), while B on the vertical axis represents the magnetic flux density or magnetic induction of the magnetic nanoparticles. The crossing point between the two axes is at zero. Here, let me explain further before you get too confused. When you were young, you may have used a permanent magnet to pull or attract a small metallic object. If it moved and it was not a permanent magnet, this object was most likely made of a ferromagnetic material such as iron, cobalt, nickel, or one of the corresponding alloys. As you approached the magnet from such a ferromagnetic object, the magnitude of the magnetic field (H) at the location of this object gradually increased until the object became sufficiently magnetized to move. This will occur when the induced magnetization of the object (represented by B) exposed to a magnetic field H will be translated into a magnetic force induced on the ferromagnetic material that would be sufficient to counteract opposite forces such as the friction force between the ferromagnetic object and the surface of the table in this particular example. In our case, that would be the various forces encountered in the bloodstream that oppose the propelling force of our submarine. Such gradual increase in the induced magnetization of our submarine is represented by the red curve going from the horizontal axis on the right side of the vertical axis, and up to a maximum value represented by the saturation magnetization level. Therefore, to achieve a maximum displacement or propelling force in the bloodstream, which proves to be critical especially at such a small scale, the magnetization of the ferromagnetic particles acting as propulsion engines in our submarine must reach the saturation magnetization level as depicted in Fig. 4.5.

However, as observed in the graph of Fig. 4.5, after the mission, i.e., once the medical intervention or the mission is completed, the magnetic field H will be removed. When such a magnetic field will be removed, the magnetization of the ferromagnetic nanoparticles will decrease from the saturation level by following the left red curve until it reaches the point of remanence on the B axis when $H = 0$. Because of this remanence, the particles will be attracted by each other after our submarine will be disintegrated. You have an example of that in Fig. 11.6 showing the nanoparticles of submarines after reaching the brain. As I mentioned earlier, except for our first 1.5 mm submarine, which had only one large engine (ferromagnetic particle) and which could be recovered because it was restricted to navigate only in larger arteries, the other submarines were made smaller in order to travel deeper in the vascular network, making recovery impossible or not practical.

The biodegradable materials used for the submarines are specially selected so that they can be rejected later by the body. But magnetic materials such as magnetic microscopic particles are not biodegradable. So these particles must be sufficiently small to be rejected by the body after the mission is completed.

In these cases, these smaller submarines rely instead on many smaller "engines" or "nanoengines" working together to provide sufficient propelling force to accomplish the mission. This is somewhat the same approach as an aircraft relying on four engines versus a mono-motor. So, when these miniature submarines made of a biodegradable material gradually fall apart, the ferromagnetic particles still acting as small magnets will attract each other. A small magnetic field will be present around each particle with the lines of magnetic field joining one side of the particle, the North Pole, to the opposite side of the particle, the South Pole. This local magnetic field is often referred to as a dipole (or magnetic dipole), making reference to the dual opposite poles. This will result in the formation of a relatively large aggregation of attracted particles where the North Pole of one particle will attract the South Pole of a neighbored particles and so on, similar to the way permanent magnets attract each others. This is referred to as dipolar interactions. From a clinical point of view, this is not good and may lead to serious complications, not to mention the inability of the body to get rid of such a large aggregation.

As depicted also in Fig. 4.5, an external magnetic field strength (referred to as the coercive force in Fig. 4.5) could be applied to demagnetize the ferromagnetic particles, but implementing such an approach in this type of platform while maintaining a safe environment for the patient would likely prove to be too complex and expensive and, therefore, not suitable if our ultimate goal is to transfer such a technology to the clinic.

4.3.4   *Superparamagnetic Propulsion Engines*

The solution relies on nanotechnology and particularly on superparamagnetic nanoparticles. Indeed, as depicted by the green curve in Fig. 4.5, superparamagnetic nanoparticles behave as ferromagnetic particles but have no residual magnetization when the magnetic field H is removed (no hysteresis in the magnetization curve). In other words, the "nanoengines" of our submarines come to a complete stop (not providing any power) when we "turn off the ignition key."

Figure 4.6 shows the fundamental difference between a ferromagnetic and a superparamagnetic particle, or perhaps I should say a ferromagnetic versus a superparamagnetic microscopic propulsion power source for our miniature submarines. Each ferromagnetic particle has several magnetic domains represented by small regions with an arrow within the larger sphere shown in Fig. 4.6 (left). A

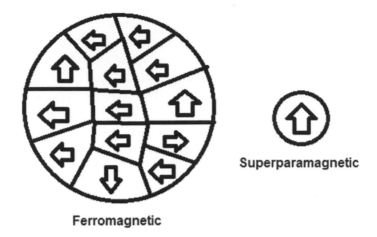

Superparamagnetic

Ferromagnetic

Figure 4.6 Multi magnetic domains in a ferromagnetic particle versus a single magnetic domain in a superparamagnetic nanoparticle.

magnetic domain is a region where the individual magnetic moments of the atoms are aligned with one another and, therefore, point to the same direction (represented by the arrows in Fig. 4.6). The magnetic moment of an atom, for instance, is a quantity that determines the torque each atom will experience in an external magnetic field. The geomagnetic field, for instance, induces a torque on the needle of a magnetic compass in such a way that it will point to the North Pole. This is the same principle here except that each individual atom acts like a magnetic compass needle where all such needles within the same magnetic domain will point to the same direction when there is no magnetic field being applied.

When we begin to apply a directional magnetic field, the magnetic moment of some of the magnetic domains in the ferromagnetic particle will begin to align with the applied directional magnetic field. Since more magnetic domains will point to the same direction, the magnetization of the ferromagnetic particle will increase following the magnetization curve depicted in Fig. 4.5. As we increase the magnetic field, more magnetic domains will align in the direction of the magnetic field. At some point, while increasing the magnetic field, all magnetic domains in the ferromagnetic particle will point to the same direction. At this moment, the ferromagnetic particle would be at saturation magnetization. In order words, our ferromagnetic power source will be at maximum regime. Increasing further the magnetic field will not increase the magnetization of the ferromagnetic particle further because all magnetic domains are already aligned in the same direction (saturation magnetization). This is like pressing the gas pedal when it is already touching the floor and cannot go further. The type of ferromagnetic materials used to implement (synthesize) the ferromagnetic particle will have an influence on the maximum regime (saturation magnetization) that our "ferromagnetic engine" can reach. For instance, the same ferromagnetic particle made of iron–cobalt will have a saturation magnetization much higher than the one made of iron oxide. This is one among many examples showing that materials sciences play a critical role in this type of projects.

But once the ferromagnetic particle is no longer exposed to the magnetic field, the magnetic domains will fluctuate (flip) towards various random directions. This could result in magnetic domains cancelling each other such that the net magnetization vector could theoretically be null. But not all magnetic domains will fluctuate and a relatively large portion will maintain their original direction, i.e., the direction when they were at saturation, hence resulting in a net vector corresponding to the residual magnetization.

For the single-domain superparamagnetic nanoparticle, the particle is sufficiently so small that when not exposed to a magnetic field, its magnetization can randomly flip direction under the influence of temperature. With a sufficiently large number of such nanoparticles in each submarine and acting as superparamagnetic nanoengines or nanopropulsion systems, the random orientations of each nanoparticle will cancel each other and, hence, result in a zero net magnetization (no residual magnetization). Because of the small overall size of our nanoparticles, not much propelling force will be produced per nanoparticle and therefore, a substantially large number of nanoparticles would be required to achieve sufficient force to propel our submarine in the bloodstream. For a superparamagnetic nanoparticle made of iron oxide, for instance, its diameter will only be approximately 10 to 20 nanometers. Therefore, each superparamagnetic propulsion nanoengine in our submarine is roughly the size of a large molecule and smaller than a virus (20 to 400 nanometers).

4.3.5   *Iron–Cobalt versus Iron Oxide Propulsion Engines*

The version of the first version of our crewless submarine (drone) and used to navigate in the carotid artery was made of a single 1.5 mm spherical chrome steel engine, while the second version of the submarines that navigated in the hepatic artery towards deep regions in the liver relied on many spherical iron–cobalt engines (nanoparticles) that were 170 nanometers in diameter. Although the engines for our submarine made of iron–cobalt superparamagnetic nanoparticles would provide the maximum "horse power" (saturation magnetization), cobalt is often regard as a toxic material for the body. This is particularly true in high dose but such a toxicity is less obvious at the quantity being used in the submarines. The use of our iron–cobalt nanoparticles made some doctors interacting with us to question if it should be injected into a patient in a future scenario. In the meantime, although we could investigate a way to prevent direct exposure of cobalt with blood by adding a protecting layer (as we did in the previous version of our submarines by adding a few nanometer thick graphite layer), we decided to go to the safer side for the next-generation submarines by adopting iron oxide superparamagnetic nanoparticles instead, which have proven to be safe since they are presently used as imaging contrast agents in magnetic resonance imaging (MRI). Notice that larger iron oxide particles are "ferrimagnetic," which means that unlike ferromagnetic particles where all domains points to the same direction when exposed to a high magnetic field, other domains will point to the opposite direction, resulting in a lower magnetization. Such iron oxide superparamagnetic nanoengines are less powerful (lower saturation magnetization) than the previous iron–cobalt nanoengines but still sufficient to navigate our submarines in the bloodstream provided that we put a sufficient quantity of these nanoengines in each submarine. Each of these nanoengines is made of molecules where each molecule has a few atoms linked together, e.g., two atoms of iron and four atoms of oxygen (Fe_2O_3) known as hematite, or with three atoms of iron and four atoms of oxygen (Fe_3O_4), which occurs in nature as the mineral magnetite, the most magnetic of all the naturally occurring minerals on Earth. As a small note, the unit typically used to describe the power of each "nanoengine" is not "horse power" but is expressed as (mass) magnetization in emu/g.

4.3.6 Generation of a Magnetic Field Sufficiently High to Bring the Propulsion Engines of the Submarine at Full Regime

One of the questions that I had initially when I began this project was how high such applied magnetic field needs to be and how much it would cost to generate it, since I knew from my previous experience in the industry that a too high cost would prevent or, in the best case, delay for a very long time the integration and the use of such a new technology. Indeed, when you do research and development (R&D) in medical technologies and many other fields, cost is always an issue and too often it is one of the reasons forcing many new technologies to stay on the laboratory shelves instead of becoming useful for the society.

The saturation magnetization of our iron oxide superparamagnetic nanoparticles is approximately 1.5 T as depicted in Fig. 4.7. To give an idea, the magnitude of the magnetic field at the Earth's surface (the one that makes the needle of a magnetic compass to point towards the North Pole) ranges from 0.000025 to 0.000065 T. This is the field of a very large magnetic dipole around the Earth, and it is similar to the one

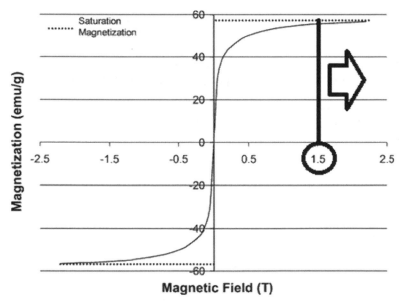

Figure 4.7 Magnetization curve of Fe_3O_4 magnetic suspension showing a saturation magnetization at approximately 1.5 T and above.

present around the superparamagnetic nanoparticles of our submarines when exposed to a magnetic field. Such a high magnetic field of 1.5 T required for our miniature submarines operating in a volume equivalent to a human adult can only be generated by a superconducting magnet. But developing such a superconducting magnet is quite expensive and takes a lot of valuable space in a hospital's operating room.

However, interestingly most modern clinical MRI scanners found in the majority of hospitals already provide a magnetic field of 1.5 T inside the tunnel of the scanner where the patient is located during the MR-imaging session. This field in the tunnel of the MRI machine is known as the B_0 field. Therefore, an obvious solution was to use the B_0 field to bring our iron oxide superparamagnetic propulsion engines (nanoparticles) at full regime (saturation magnetization).

So going back to Fig. 2.2, you will see the presence of a clinical MRI scanner in our interventional room. This scanner is used not only for imaging the interior of patients as typically done in hospitals but also to bring the nanoengines (superparamagnetic nanoparticles) of our submarines when operating in the bloodstream at full regime (saturation magnetization) anywhere and at any depth in the body. The same MRI scanner is also used to drive or navigate the tiny submarines along a preplanned trajectory in the blood vessels. How this is done will be explained later in this book.

Indeed, it is worth mentioning that the B_0 field is a uniform (homogeneous) magnetic field. In other words, the magnitude of the magnetic field is the same anywhere in the central volume of the tunnel of the MRI machine where imaging and the navigation of our submarines occur. Such a magnitude will be 1.5 or 3 T for a 1.5 or 3 T clinical MRI scanner, respectively. Because there is no variation of the magnitude of the magnetic field over distance, the latter referred to as a magnetic gradient, the submarine will not move (unless there is another force involved such as the one caused by the blood flow). In other words, it is as if you press the gas pedal (accelerator) of your car to the floor to obtain the full regime (if saturation magnetization is achieved, which is the case when using the B_0 field of a clinical MRI scanner) with the clutch or transmission in the neutral position. To move and drive towards a specific direction in order to navigate in the bloodstream, you will need a 3D directional magnetic gradient. Again, this will be explained in detail later in this book.

Another fact about the B_0 field is that the lines of magnetic field are oriented towards the back end of the MRI scanner and along the longitudinal axis of the tunnel (known as the z-axis of the MRI scanner). If our submarine had a shape similar to the submarine in the movie, it would have always been oriented along the z-axis (the nose of the submarine would always point towards the far end of the MRI tunnel), independent of the direction of the blood vessels. In other words, the submarine would behave like a miniature magnetic compass needle that always points to the North (towards the far end of the tunnel of the MRI scanner) even when we turn the compass (change the direction of travel). This is one main reason that explains the spherical shape (isotropic instead of an anisotropic shape) of our submarines since they will always looks similar with the same amount of surface being exposed to the blood flow in order to keep the drag force constant whatever the direction of travel would be in the blood vessels.

I should also mention that we did the first navigation experiments using a 1.5 T MRI scanner. The clinical MRI scanner depicted in the interventional room in Fig. 2.2 has a B_0 of 3 T, sufficient to bring at full regime our iron–cobalt nanopropulsion units anywhere and at any depth inside a human body, if we decide to use them at some point instead of the less powerful iron oxide version.

4.4   At the Pilot Seat (Scene 00:21:37–00:21:52)

In this scene, you see where the pilot sits with the instrument panels in front of him, which appears relatively simple. Our submarines are piloted (navigated) from the control room, typically in the autopilot mode. Such a mode is programmed to navigate each submarine along a preplotted route in the bloodstream. We also have a computer display but larger and typically more complex than the one shown in the movie. Several computer-generated windows can be displayed on the "pilot display" located in the control room situated next to the interventional room (see Fig. 2.2). An example of a computer screen (at the pilot seat) during navigation is shown in Figure 4.8. Although the computer interface is being remodeled to better respond to the needs of medical specialists for specific interventions, the example

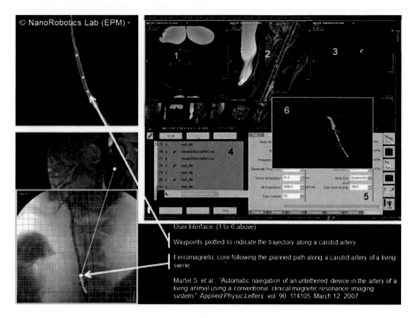

Figure 4.8 An example of a computer display showing various details during the navigation phase. From *International Journal of Robotics Research* 28(9):1169–1182, copyright © 2009, with permission from SAGE Publications.

depicted in Figure 4.8 that was used for the first navigation in the carotid artery gives an overall idea of the type of information that can be displayed.

On the top left section, you can see a reconstructed image of the carotid artery with the pre-programmed waypoints being plotted on top. This reconstruction of the image of the carotid artery is done from a process known as segmentation of the real images. The waypoints represented here as dots indicate to the automatic control navigation software the course to be followed by the submarine. Just below, you see the tracking signal generated by the submarine and further below, you see the position of the submarine being tracked in the carotid artery. MR-image and X-ray image (the latter showing the spinal cord) are superposed and aligned to provide better information to the "pilot." The fusion of images from more than one medical imaging technology is known as multimodal imaging. The alignment of these images, including the tracking position of the submarine, is known as image registration.

In Fig. 4.8 (top), you see three displays labeled 1, 2, and 3, which show actual MR-images of the physiological regions being navigated. The two parallel lines superposed on the images are used to measure various features such as the sizes and the angles of the blood vessels to be navigated. The other displays labeled 4 and 5 are two windows among many that are used for setting various parameters for imaging, real-time tracking, and navigation. The last display labeled 6 shows the tracked displacement signal of the submarine.

In the new implementation, two main computer displays are used instead of one. Everything related to MR-imaging is displayed on one computer screen, normally the one dedicated to MR-imaging, while the other computer display is dedicated for navigation. The navigation computer receives from the imaging computer the image data, which in turn are used to perform the navigation task.

At the end of the scene starting at 00:21:47, you see that a joystick is used to maneuver the submarine. Interestingly, we considered using a joystick and we actually used at some point a joystick similar to the one depicted in the movie. However, it was abandoned later when we saw that performing computer-based automatic navigation in the bloodstream prove to be much faster and reliable.

4.5 Onboard Weapon (Scene 00:21:52–00:23:20)

In this scene, Grant is joining Cora, who is busy testing the laser that will be used to destroy the clot in the brain. Well as you may expect, although carrying such a laser onboard our submarines would be a nice option, we do not know yet how to miniaturize a laser gun as the one shown in the movie. Instead, our submarine carried in the past chemical weapons known to be effective at destroying the type of tumors that we targeted. An example is doxorubicin. Other types of weapons against cancer could also be transported and the list of such potential weapons can be quite long. The list includes not only therapeutic agents but also the ones for radiotherapy and immunotherapy, to name but a couple of examples.

The other interesting fact in this particular scene is when Cora mentions that the laser can be regulated to one millionth of a millimeter, which corresponds to 1 nanometer! This is well before that nanotechnology and the scale of 1 nanometer were popular topics in the scientific community. This is probably another coincidence.

4.6 Name of the Submarine (Scene 00:24:50–00:25:00)

In this scene, we learn that the name of the submarine is *Proteus*. The name of our first submarine, the one that like *Proteus* was injected into the carotid artery, was called MR-Sub for Magnetic Resonance Submarine, making reference to a submarine being propelled by a magnetic resonance imaging machine. The following small versions took other names. The one in Figure 4.1, for instance, was referred to as TMMC having four characters as for the name CMDF in the movie. TMMC stands for therapeutic magnetic micro carriers.

5

Miniaturization

I N THE FOLLOWING scenes of the movie, we witness the submarine being miniaturized prior to being injected into the carotid artery. In this chapter, we discuss the miniaturization aspect and compare it to the process seen in the movie.

5.1 Phase 1 of the Miniaturization Process (Scene 00:25:05–00:28:10)

It is in this scene that the first phase of the miniaturization process is initiated. Here, the submarine with its crew is miniaturized to an initial size prior to be miniaturized further to a size adequate to navigate in the arterial network.

Although the versions of our submarines such as the ones depicted in Figs. 4.1 and 4.3 cannot be miniaturized and as such must be implemented (constructed) at an overall size adequate for navigating in the arterial network, we developed one version that can be miniaturized. You may be surprised and probably skeptic to hear that making a real submarine that can be miniaturized can in fact be a reality. But to convince you, I will describe in the following lines how the miniaturization of a submarine was done by our group in 2010.

A Microscopic Submarine in My Blood: Science Based on Fantastic Voyage
Sylvain Martel
Copyright © 2016 Pan Stanford Publishing Pte. Ltd.
ISBN 978-981-4745-78-9 (Hardcover), 978-981-4745-79-6 (eBook)
www.panstanford.com

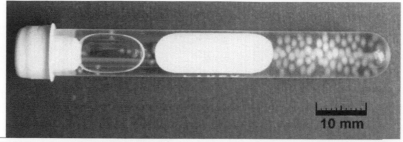

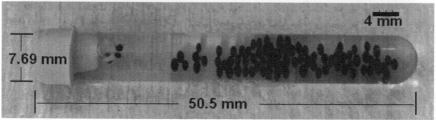

Figure 5.1 Photograph of the PNIPA submarines. The top photograph shows PNIPA submarines without magnetic nanoparticles while the bottom photograph shows the same PNIPA submarines but loaded with magnetic nanoparticles used as magnetic nanopropulsion systems. With the magnetic nanoparticles, the PNIPA submarines change color from white to black without requiring a paint job as for *Proteus*! Images from *Advanced Robotics* 25:1049–1067, copyright © 2011, with permission from Taylor & Francis.

These shrinkable submarines depicted in Fig. 5.1 are referred to here as PNIPA submarines. They are part of a new class of vessels capable of navigating the arterial network and changing their dimensions. They are referred to as hydrogel-based magnetic submarines. PNIPA stands for poly(*N*-isopropylacrylamide) and it is a thermo-sensitive hydrogel. It is the material used to make the structure of this special class of navigable submarines.

A hydrogel is a network of polymer chains that are hydrophilic and can absorb a relatively large quantity of water. PNIPA hydrogel submarines are able to reduce their overall size in response to an elevation of temperature over a transition temperature referred to as the lower critical solution temperature (LCST). The LCST of PNIPA hydrogel is 33°C, slightly lower than the temperature inside the human body at 37°C but can be adjusted at or slightly above the human body temperature by modifying the ratio of hydrophilic versus hydrophobic

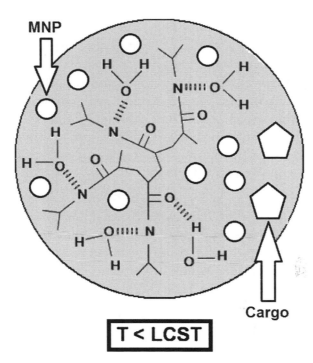

MNP

Cargo

T < LCST

Figure 5.2 Simplified diagram of a navigable PNIPA submarine before being miniaturized. Adapted from *Advanced Robotics* 25:1049–1067, copyright © 2011, with permission from Taylor & Francis.

components. Figure 5.2 depicts a simplified schematic of the PNIPA magnetic submarine before miniaturization when the surrounding temperature, *T*, is less than the LCST. The diagram shows the PNIPA molecular structure with the magnetic nanoparticles (MNPs) used as propelling nanoengines next to the payloads. The payloads can be anything, from therapeutic drug molecules to microscopic crew members.

Several scenarios can be envisioned to initiate a miniaturization phase similar to the one in the scene of the movie. A platform under the submarine similar to the one in the movie can be used, which acts as a heating plate. Another method to elevate the temperature inside the submarine above the LCST to initiate the miniaturization phase could be through hyperthermia using the magnetic nanoparticles not only as propelling power source but also as heating sources. This can be done by modulating the applied magnetic field at higher frequencies.

For instance, as mentioned previously, superparamagnetic nanoparticles of Fe_3O_4 with a diameter less than 20 nanometers (nm) are typically used to serve as the nanopropulsion units for our submarines. Because they are superparamagnetic, the orientation of their magnetic moment (remember the magnetic compass needle) continuously changes due to thermal agitation. An external magnetic field with sufficient energy to overcome the thermal energy barrier of the nanoparticles will rotate the magnetic moments of these nanoparticles until they align with the direction of the applied magnetic field. Then, when the magnetic field is removed, each magnetic moment does not relax immediately but takes a short time before returning to a random orientation. This is known in nanotechnology as the Néel relaxation mechanism. It is during this relaxation period that the magnetic field energy of each nanoparticle is released in the form of heat. We typically repeat this process at approximately 100,000 times or more per second (100 kHz) to create sufficient heat using a special platform that we call the hyperthermia station (see the location of the hyperthermia station in the interventional room in Fig. 2.2) which is capable of generating and modulating a suitable magnetic field to do just that. A frequency of 100 kHz has been selected since it prevents harmful physiological responses from a human patient while such a frequency allows the electromagnetic energy being generated to penetrate approximately 10 cm inside the human body, which will prove useful when we will need to access regions of the brain as described later in this book.

These previous scenarios would miniaturize the submarines before they are injected into the artery. Although this would resemble more the scenario of the movie, it will probably not be the best approach to consider in our particular case since unlike in the movie, we can already build submarines that have an initial overall size suitable for navigating in the bloodstream.

A more logical approach in our particular case is to initiate the miniaturization while the submarine is in the body. Such miniaturization process could be initiated as soon as the submarine enters the artery since the internal body temperature would be higher than the LCST. This could be done to release drug cargos for a crewless PNIPA submarine or, in the case of a crewed submarine, to

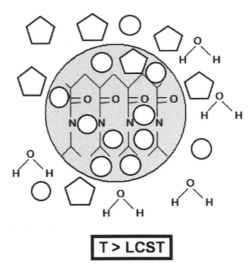

Figure 5.3 Simplified schematic representing the PNIPA submarine after the miniaturization process; as depicted in the diagram, water molecules (H_2O) that were part of the structure of the submarine are detached and expelled, which results in a shrinkage of the submarine. Some magnetic nanoparticles represented by the white circles and the cargos which could be microscopic crew members represented by pentagons are expelled during the process. Adapted from *Advanced Robotics* 25:1049–1067, copyright © 2011, with permission from Taylor & Francis.

allow microscopic crew members to exit the PNIPA submarine. This is shown schematically in Fig. 5.3.

Relying on the body temperature to initiate the miniaturization process means that our crewed submarines would have a limited amount of time to reach the target. Figure 5.4 shows three miniaturization curves of our initial PNIPA submarines.

As depicted in Fig. 5.4, the present version of the PNIPA submarine can be miniaturized down to approximately 30% or one-third its original size. Although this level of miniaturization is not as high as the one achieved in the movie with the submarine *Proteus*, even with the miniaturization level achieved during phase 1 of the miniaturization process of *Proteus*, it still demonstrates that a real submarine designed to navigate in the arterial network can be miniaturized considerably.

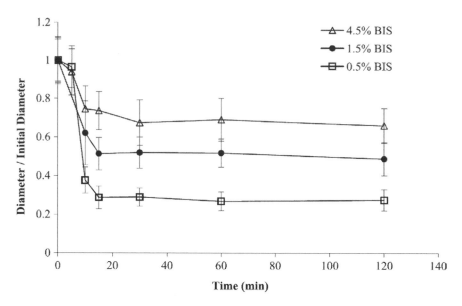

Figure 5.4 Miniaturization over time of the initial PNIPA submarines synthesized (constructed) with different cross-linker (BIS) concentrations. From *Advanced Robotics* 25:1049–1067, copyright © 2011, with permission from Taylor & Francis.

As shown in Fig. 5.4, changing the cross-linker (BIS) concentrations for the structure of the PNIPA submarine will have a direct impact on the level of miniaturization that can be achieved. A cross-linker is a chemical reagent designed to covalently interact with the molecules of interest, resulting in conjugations. In simpler words, it is a molecular piece that can link together two molecular pieces (molecules), allowing one to build a larger structure. It must be compatible with the molecular pieces that needed to be assembled (i.e., choosing the right screw or mechanical joint that will fit in the sockets of the pieces to be assembled). The BIS (*N,N'*-methylenebisacrylamide) cross-linker is one example that is used to assemble the structure of our PNIPA submarines.

Then looking at Fig. 5.4, you will conclude that we should use less BIS cross-linkers to build our submarine in order to achieve a higher level of miniaturization. But as expected, by using a lower concentration of cross-linkers, the mechanical structure of your PNIPA or hydrogel submarine becomes less durable. By analogy, it is similar to using fewer screws to mount a structure; as you will expect, it may not be as solid as the version with a larger number of screws. So a compromise of 3% of cross-linkers was initially used to build our PNIPA submarines.

Another issue is the time available for the submarine to reach the target before the payloads or the microscopic crew members are expelled from the PNIPA submarine. According to the graph in Fig. 5.4, the total miniaturization process takes approximately 15 minutes. This is much longer than the total time of 1 minute and 19 seconds (first phase (27:20–28:09) = 49 seconds + final miniaturization phase (34:08–34:38) = 30 seconds) required to miniaturize *Proteus*.

Another concern using the body temperature as a trigger for the miniaturization process is that some payloads or microscopic crew members could be expelled during the initial 15 minutes while traveling towards the final destination. As such, one could envision increasing the LCST slightly above 37°C and use the hyperthermia approach discussed earlier to allow the microscopic crew members to exit. Besides the fact that this process may take many precious additional minutes to accomplish and which is a real concern because of the limited survival time of our crew members in the body, this would increase further the internal temperature of the submarine, which may reduce further the lifetime of our crew members.

5.2   Phases 2 and 3 of the Miniaturization Process (Scene 00:28:10–00:34:38)

In this scene, the submarine inside a large mockup of a syringe is further miniaturized to the size appropriate for navigating in the bloodstream. You can see that during the miniaturization phases, the submarine and the crew members of the *Proteus* are placed on top of a special platform with several hexagons on its surface (for example, one among several shots in the movie where you can see this is at 00:33:50) that are used for the miniaturization process. This is similar to our setup except that the platform is smaller and the hexagons are replaced with several dishes (here with a round shape but they could be hexagonal as well) as shown in Fig. 5.5.

The platform depicted in Fig. 5.5 is used to miniaturize the overall size of the "crew" prior to the injection into the body. Such a process

Figure 5.5 Platform used to reduce the overall size of the crew prior to the injection. It is common practice to give the first name of one of the members of the team to a particular system that has been developed in the lab. Here this particular platform was given the name "Samira."

is required because the number of "crew members" required for fighting cancer largely exceed the number of crew members in the movie *Fantastic Voyage*. Indeed, we concluded that much more than five crew members shown in the movie would be required. In fact, our preliminary experiments suggested that a total exceeding 100 million crew members would be required per mission. Initially, these so-called "microscopic crew members" are maintained in bottles as shown in Fig. 5.6, waiting to be called for the next mission. We cannot see the microscopic crew members in the bottles because they are too small but you will notice the foggy liquid in the bottles. This foggy aspect is due to the presence of a large population of these microscopic crew members.

When in a bottle, these microscopic crew members are dispersed in a relatively large volume of liquid. Injecting these crew members would lead to a huge quantity of fluid being also injected at the same time. This is somewhat the same issue as in the movie when *Proteus* with its crew members is submerged in the oversized syringe. It is obvious that they could not inject this huge quantity of liquid contained in

Figure 5.6 The bottles containing the dispersed (not concentrated) microscopic "crew members."

this larger syringe into the carotid artery of the patient. They had to decrease the volume of liquid necessary to inject the crew members by proceeding to the final miniaturization process.

Since in our case we cannot miniaturize the bottles, we reduce instead the overall size (volume) of the population of these microscopic crew members. To do that, we extract a quantity of fluid from a bottle that we know from the initial concentration would have the right number of crew members needed for the mission. We do that using a syringe that looks exactly as the one in the movie but after, and not before, it has been miniaturized. Then the extracted fluid containing our microscopic crew members is placed on the dishes of the platform depicted in Fig. 5.5 using the same syringe. The whole quantity of fluid is distributed between several dishes to accelerate the "miniaturization process." When all of the liquid has been deposited in each dish, the power is switched on from the power supplies depicted in Fig. 5.7.

When the power supplies in Fig. 5.7 are switched on, the electromagnetic units are activated. Each electromagnetic unit consists of a coil of electrical wires with a ferromagnetic core inside. The ferromagnetic core is used to concentrate the magnetic

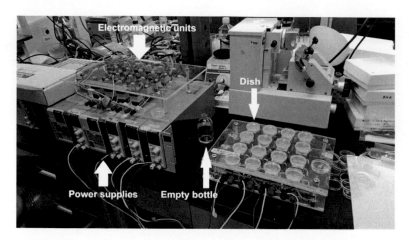

Figure 5.7 Photograph of the setup showing the dishes on top of the electromagnetic units (some without the dishes are shown on the top left section of the photograph), each dish containing a fraction of the whole population of "microscopic crew members," with the power supplies providing the power to the electromagnetic units.

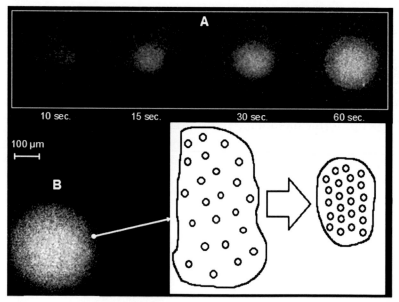

Figure 5.8 (A) Formation of an aggregate of the microscopic crew members initially dispersed in a large volume and resulting in a smaller overall volume of the crew 60 seconds later. (B) Final smaller (concentrated) crew in a much smaller overall volume. The process is also represented schematically (bottom-right) where a crew initially in a large volume shrinks to a much smaller size when the electromagnetic unit is activated.

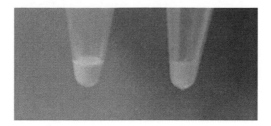

Figure 5.9 Two crews ready for injection, each crew has approximately 100 million microscopic members still alive and ready for the mission. At the end of the process, the 100 million microscopic crew members that were in the bottles depicted in Fig. 5.6 are now reduced to a final overall size equivalent to a few drops of water.

field to the center of each dish containing a portion of the whole population of microscopic crew members that will be injected. Once the electromagnetic units are activated, the size of the population of microscopic crew members decreases. An example but with only a few thousands microscopic crew members is depicted in Fig. 5.8.

With several electromagnetic stations (dishes) operating in parallel, shrinking the overall size of a crew of approximately 100 million microscopic members takes only a few minutes. A single station would have taken much longer to complete. Once the process is complete, the microscopic crew members are extracted from each dish using a syringe similar to the one used in the movie before being put temporarily in a small container while waiting for the injection as shown in Fig. 5.9, which shows a population of approximately 100 million microscopic crew members appearing as a very small volume of concentrated white fluid.

6

Injection

THIS CHAPTER IS dedicated to the injection process. You will see that it can be as simple as inserting the needle of a syringe into the patient as shown in the movie, but it can also be much more complicate than the injection process depicted in the part of the movie that begins at 00:37:09 and finishes at 00:38:12.

6.1 Beginning of Injection with a Syringe (Scene 00:37:09–00:37:41)

In this scene, we see a syringe filled with a fluid containing the submarine *Proteus* with the five crew members onboard, approaching the injection site indicated by an "X" printed on the skin of the patient. We do not indicate the location of the injection site by printing an "X" on the skin of the patient.

6.1.1 *Injection Being Done Close to the Target*

We do use a simple syringe similar to the one shown in this scene, but such a syringe alone is typically used to inject the "microscopic crew members" without any submarines. Since no submarines are

A Microscopic Submarine in My Blood: Science Based on Fantastic Voyage
Sylvain Martel
Copyright © 2016 Pan Stanford Publishing Pte. Ltd.
ISBN 978-981-4745-78-9 (Hardcover), 978-981-4745-79-6 (eBook)
www.panstanford.com

typically involved when the syringe is used, the injection must be performed very close to the targeted location that needs to be treated in order to reduce the distance sufficiently to allow the microscopic crew members to swim from the injection site to the targeted location. For a solid tumor, for instance, we will do what is being referred to as a peritumoral injection. The injection will typically be done at a maximum of a few millimeters from the region that needs treatment. A few millimeters sound very small but at the scale of our microscopic crew members, this is a quite long distance. Indeed, with an overall size of 1 to 2 micrometers (like for humans, some are bigger and some are smaller than others but overall they all remain within a range of possible heights and weights), each millimeter would correspond to 500 to 1000 body lengths. Therefore, each millimeter would correspond for a miniaturized human, roughly to a distance of 1 kilometer. If you are too far, this can quickly translate into a huge distance to swim and you may need a boat (or a submarine). The main reason that each of our crew members has an overall size between 1 to 2 micrometers only is explained later in this chapter.

An example of approximately 100,000,000 microscopic crew members (see Fig. 5.9) being injected with a syringe in the periphery of a solid colorectal human tumor grown (implanted) in a mouse (such implanted tumor is often referred to as a xenograft) is depicted in Fig. 6.1. In this example, one crew of approximately 100 million microscopic crew members shown as a white fluid in one of the two vials depicted in Fig. 5.9 is extracted using the same syringe used for the injection.

Once injected, as shown in Fig. 6.2, the members of this large group will swim towards the tumor. It will take less than 30 minutes and most likely just a few minutes to cover the distance required to reach the small tumor depicted in Fig. 6.2. This is because our crew members can swim much faster than the crew members of *Proteus*. The fastest Olympic human swimmers can swim at a maximum of one body length per second. On average, our microscopic crew members can swim between 100–200 body lengths per second and a small percentage of them can even swim faster than that. To put that in perspective, an average car is approximately 4.5 meters long. So, 100 body lengths per second would correspond to a car running at 1660 km per hour and

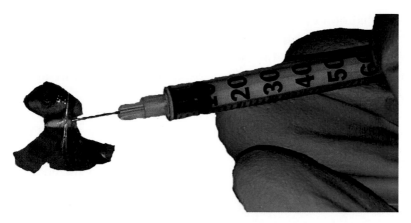

Figure 6.1 Photograph showing 100 million microscopic crew members being injected (peritumoral injection) using a syringe similar to the one used in the movie. The skin in the region of the injection has been removed here to show the population of crew members. Such population is so large that it is visible with the naked eyes and appears as a large white spot in the red physiological tissue at the end of the needle.

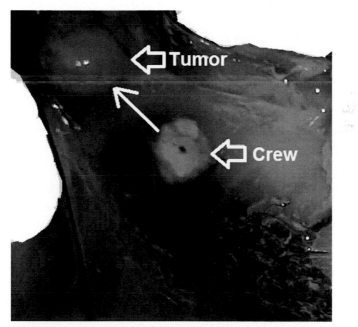

Figure 6.2 Photograph showing 100 million "armed microscopic crew members" swimming together towards a colorectal tumor. When the tumor is larger than the one shown here (diameter of 1 cm), a higher number of armed microscopic swimmers will most likely be injected.

double that of it can run at 200 body lengths per second. Think this is fast; now consider that this is done under water.

Similar to humans, the microscopic crew members prefer to swim at a comfortable temperature such as at room temperature at around 23°C. But when they swim in the blood, which has a much higher temperature at 37°C, the performance of our microscopic swimmers gradually decreases as you would expect for a miniaturized Olympic human swimmer under the same conditions. We found that after approximately 30 to 40 minutes, our microscopic crew members could not swim anymore. So, if they could not reach the targeted location within 30 minutes, then the use of a submarine to transport them closer to their final destination would be the way to go. Indeed, peritumoral injections are possible for some type of tumors located in regions that are more accessible to perform the injections. But not all regions are easily accessible. The liver and the brain are two examples where tumoral injections are not practical. In these cases, the use of one or more submarines would be required. For the former, the submarines would travel through the hepatic artery, while for the latter, the crewed submarines would travel through the carotid artery as shown in the scenario of the movie.

Another question that you may have is how they know in which direction they should swim. The answer is simple. They know because we indicate to them which direction they should swim in order to reach the tumor. Then you probably want to ask how. I will tell you this later in the book.

Another question that a lot of people ask me is why not injecting them directly into the tumor. The answer is simple: because of what is known as the tumor interstitial fluid pressure (TIFP). Indeed, if you try to inject directly into a tumor, the microscopic crew members or the miniature submarines would be expelled from the tumor at a fast rate due to this TIFP. So the best way to access it is through what is called the angiogenic blood network. The angiogenic network consists of blood vessels grown by the tumor in order to bring the nutrients and oxygen required to grow. To reach the tumor, you can travel through these angiogenic blood vessels that have a diameter of only a few micrometers, the interstitial spaces which are spaces between cells that also offer physiological routes up to a maximum of a few micrometers, and small openings of less than 2 micrometers in diameter that lead to the interior of the tumor. The presence of these openings is due to the fact that tumor blood vessels are leaky, i.e., that unlike normal blood vessels, they have holes that can be as large as 2 micrometers. These are potential physiological routes available for our microscopic crew members to swim towards the interior of the tumor. Because the "entry doors" to the interior of the tumor do not exceed 2 micrometers, the size of each of our microscopic swimmer must not exceed 2 micrometers.

6.1.2 *Injection Being Done Far from the Target*

As mentioned earlier, it is often impossible to inject the microscopic crew members close enough to the targeted physiological region. In such a case, the crewed or crewless submarines such as the ones described in Chapters 4 and 5 must be injected into a larger blood vessel, typically an artery. In many instances, a simple syringe such as the one shown in the movie could be sufficient. In other instances, the syringe is connected to a microcatheter inserted into the artery. If maximum navigation accuracy is required especially for a preplanned trajectory characterized by several directional changes in a relatively short distance, then a navigation technique dubbed magnetic resonance navigation (MRN) is used. This method has been executed to navigate the first untethered object in the blood vessel of a living animal. MRN has been pioneered by our group and has been used many times in previous navigation experiments performed in both the carotid and the hepatic artery so far. How it works will be explained later in this book.

Because MRN is a very precise navigation method, the injection system becomes more complicate compared to the use of a simple syringe as depicted in the movie. Although several configurations are possible, Fig. 6.3 depicts one configuration that is presently under investigation.

As shown in Fig. 6.3, a balloon close to the injection tip introduced in the artery is inflated to reduce the blood flow rate in the artery. Controlling the blood flow rate will control the velocity of the microscopic submarines. Reducing the blood flow rate is also required to provide more time for the microscopic submarines to steer towards

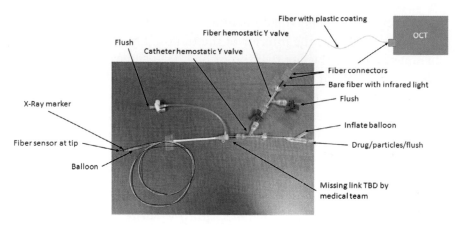

Figure 6.3 Example of an injection system for precise navigation in the arterial network.

the targeted branch at each blood vessel bifurcation. In Fig. 6.3 (right), you can see the inlet for inflating the balloon. Typically the air pressure going through this inlet is synchronized by the navigation computer. There is also an optical fiber at the tip of the catheter used to inject the microscopic submarines. This optical fiber is connected to an optical coherence tomography (OCT) system. It is used to detect the microscopic submarines at the exit of the tip and to measure the velocity of the blood. This information is critical to evaluate the time to initiate the next steering phase (new directional propulsion) of the microscopic submarines at the earliest possible time. The OCT, unlike other techniques, allows fast measurements since the submarines typically take a tiny fraction of a second to travel between two successive bifurcations. In the navy, we refer to this as dead reckoning (DR). DR is the process of calculating the current position by using a previously determined position (such as the OCT (or other methods) detection time at the tip of the catheter), based on the estimated velocity over the elapsed time and course.

   Without going into too much technical details, the last part worth mentioning here about the injection system depicted in Fig. 6.3 is the inlet for the microscopic submarines (identified as drug/particles/ flush). This is where the microscopic submarines enter the injection system. The injection is also triggered by the navigation computer in order to synchronize the whole navigation process.

At the end of this scene of the movie, we see that it is not a doctor or a human who manipulates the syringe. In fact, the piston used to inject the fluid in the syringe is performed by a machine. The same is true in our platform because the right quantity of fluid, the time to initiate each injection, and velocity at which the fluid containing the microscopic submarines must be expelled from the injection catheter are some of the parameters that must be precisely controlled and coordinated by a computer since such level of control is beyond what a human can achieve.

Right at the end, we can see in the scene (00:37:35–00:37:41), the position of *Proteus* being tracked in the syringe during the injection process. We do not track our microscopic submarines in the injector since this is not necessary, but as mentioned earlier, we detect them when they exit the tip of the release (injection) microcatheter.

6.2 Velocity of the Submarine during the Injection (Scene 00:37:41–00:38:12)

In this scene, we see *Proteus* traveling inside the syringe at a very fast speed. We can almost feel how fast it goes when later in the scene we see how the crew react and hear the sounds of the submarine traveling at high velocity.

This is another fact in the movie that agrees with the way we do it. The main reason that we inject the microscopic submarines at very high velocities is to prevent a blockage in the injection system. Although there is no real risk of a blockage by injecting only one microscopic submarine as shown in the movie, injecting several microscopic submarines at the same time could cause a blockage in the injection microcatheter because of a possible formation of a relatively large aggregation of submarines due to the force of attraction between neighboring microscopic submarines. Before I explain this in a bit more detail, let me tell you first the reason for the need to inject more than one microscopic submarine.

First, because we do not have a powerful laser that can be miniaturized as shown in the movie, we estimated based on preliminary experimental results that at least 100,000,000 armed crew members could be required to accomplish the mission, not destroying a clot in the brain as shown in the movie, but a solid tumor that could also be in the brain. Depending on the size and how aggressive the tumor is, even more armed crew members could be needed. But let assume for now that we need 100,000,000 crew members, each carrying a tiny chemical weapon (chemotherapeutic molecules) instead of a tiny laser gun in order to deliver all together sufficient destroying power to eliminate the tumor since one armed crew member alone would not do much harm, because of its size.

We mentioned earlier in this book that the microscopic submarines could be as small as approximately 10 micrometers across, but that the ideal overall diameter of a crewed submarine intended for a mission in the human body would typically be between approximately 150 and 200 micrometers. It should be sufficiently small to navigate in the arteries and sufficiently large to stuck in narrower blood vessels sufficiently close to the targeted region such as the solid tumor, so that the microscopic crew members could swim beyond the accessible range of the submarine and towards the target. In the medical jargon, such object blocking the blood vessel is known to as embolization. We used, for instance, the microscopic submarine depicted in Fig. 4.1 to perform chemoembolization in deep regions in the liver following navigation in the hepatic artery. Here, chemoembolization refers to an embolization but the one in which chemotherapeutic molecules are released from each microscopic submarine through a pre-programmed biodegradation (done by chemical synthesis) of its structure, which in this case was made of a biocompatible and biodegradable polymer known as poly(lactic-co-glycolic acid) (PLGA).

Such an embolization of the submarine is essential particularly when we release crew members. Indeed, to indicate or communicate the swimming direction to our microscopic crew members to allow them to reach a targeted site such as a tumor, we must transfer the patient from the MRI scanner to another platform known as the magnetotaxis platform. This platform is unique to the NanoRobotics Lab and you will not find such a platform anywhere else, at least in a very near future. By doing so, the submarines are no longer exposed to the B_0 field of the MRI scanner, and therefore all submarines lose their power (loss of magnetization of the superparamagnetic nanoparticles). Therefore, without embolization, our submarines would drift out of control in the systemic blood circulatory network. By analogy, the embolization process can be seen as dropping the anchor.

So if we assume a diameter approximately twice the thickness of a human hair while considering the overall size of our microscopic crew members, dividing the volume of the submarine ($V = \frac{4}{3}r^3$, where r is the radius) by the volume occupied by each of our microscopic crew members you get the maximum theoretical number of passengers of armed crew members that each of our microscopic submarine could transport. You might be surprised when finding how large each crew can be and how many submarines would need to be injected per mission. Compared to the invasion of Normandy (Fig. 6.4) during World War II (D-day), known to be the largest seaborne invasion in history and which involved the landing of 24,000 soldiers, the number of microscopic soldiers required in our case to eliminate a tumor may indeed suggest a war ahead.

Figure 6.4 Image of the invasion of Normandy, which represents well the tactic used by our crewed vessels releasing microscopic armed soldiers to fight cancer cells. Image taken from https://en.wikipedia.org/wiki/Normandy_landings#Allied_order_of_battle.

So now that we know that many microscopic submarines must be injected in order to win the war against cancer, the question is why we should inject each microscopic submarine at a high velocity such as in the movie. The answer is related to the dipolar attractive force between the microscopic submarines. At the time of the injection, the patient lies in the tunnel of the clinical MRI scanner. One reason is that we need to have the nanopropulsion units ready and at full regime (superparamagnetic nanoparticles at saturation magnetization) in order to operate as soon as the microscopic submarines enter the blood vessels, otherwise the submarines will drift out of control.

When the superparamagnetic microscopic submarines approach the tip of the injection system located in the tunnel of the MRI scanner, their magnetization increases gradually as they approach the B_0 magnetic field inside the tunnel of the scanner (see the graph in Fig. 4.5) until they reach saturation magnetization once inside the tunnel of the scanner. Such a magnetic field near the entrance of the tunnel will increase sharply. Such fast increase of the magnetic field corresponds to a high magnetic gradient that will accelerate the superparamagnetic submarines towards the tunnel, especially very near the entrance of the tunnel where the magnetization of the superparamagnetic microscopic submarines would be higher for the same directional gradient. This field outside the tunnel of the MRI scanner is part of the fringe field (the magnetic field emanating outside the scanner and which is generated by the superconducting magnet used to produce the B_0 field). When inside the tunnel in the B_0 field, there will be no gradient since the field becomes uniform, and therefore no magnetic force will be present to displace the microscale submarines. When this occurs, there will be a dipole magnetic field around each magnetic nanoparticle acting as nanoengines for the submarines. If the submarines are not moving sufficiently fast and the distances between neighboring submarines are sufficiently small, then they will be attracted to each other to form aggregations of submarines. If this accumulated mass of microscopic submarines becomes as large (or larger) than the orifice of the injector, then they can jam the injector and prevent the injection of the submarines. The best way to avoid that is to maintain a sufficiently high velocity of the injected fluid as observed in the movie. But if the injection is performed too fast, then the submarines could travel a too long distance from the tip, leaving not enough time to steer them to the appropriate channel at the next blood vessel bifurcation, resulting in a scenario similar to the one observed later in the movie where *Proteus* ends up in the wrong blood vessel soon after being injected into the carotid artery.

7

Traveling in the Artery

IN THE FOLLOWING scenes following the injection, the submarine *Proteus* is seen for the first time traveling in the carotid artery. There are a few interesting facts in the scenes that are worth mentioning. This is what you will discover in this chapter.

7.1 Size of *Proteus* (Scene 00:39:25–00:39:35)

In this scene, we see red cells drifting around *Proteus*. Knowing that a red cell has an approximate length between 4 and 6 micrometers, then from what we see in this scene, we can estimate that the submarine has a length of approximately 30 micrometers and a width of approximately 20 micrometers. Those are rough estimates but still give an idea of the overall dimensions of the submarine. These dimensions are within the range of various submarine sizes developed by our team. But the problem with smaller submarines is that they can only carry a few crew members, which is fine with the five crew members in the movie but problematic in our case. As mentioned in Chapter 5, larger submarines (150–200 micrometers) capable of navigating in the arteries while carrying more crew members are preferable in our

A Microscopic Submarine in My Blood: Science Based on Fantastic Voyage
Sylvain Martel
Copyright © 2016 Pan Stanford Publishing Pte. Ltd.
ISBN 978-981-4745-78-9 (Hardcover), 978-981-4745-79-6 (eBook)
www.panstanford.com

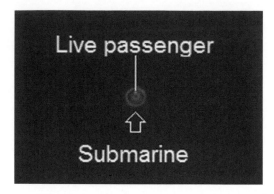

Figure 7.1 Optical microscopy photograph of a single-passenger submarine; the diameter of the submarine is approximately 10 micrometers. The live passenger appearing as a small dot can barely be seen inside the tiny submarine.

particular applications. Figure 7.1 depicts a 10 micrometer submarine carrying only one of our microscopic crew members, just to show you that building submarines even smaller than *Proteus* in its final miniaturization state is possible.

7.2 One Hundred Thousand Miles Long
(Scene 00:40:22–00:40:24)

In this very short scene, Grant learns that the blood circulatory network is 100,000 miles long. This is bit longer than what I mentioned in the Prologue (100,000 kilometers) but it is still in the range. But the scene shows a big difference with the vocabulary typically used by the scientific community, including the one in the USA. Indeed, it is surprising to hear a scientist not using the metric system as it is common practice when scientists communicate with each other. But Grant is not a scientist, which may explain the reason that the word "mile" was used instead. Grant is an American, and the USA still uses the miles even 50 years after the release of the movie *Fantastic Voyage*. So nothing has changed in this respect in the past 50 years. But in Canada, where the new "fantastic voyages" occur, the kilometer is being used today by all Canadians. But at the time of the scene, i.e., 50 years ago, people in Montréal and across the country had not yet started using the metric system.

7.3   Speed of the Submarine (Scene 00:41:29–00:41:32)

In this very short scene, we learned that *Proteus* is traveling in the carotid artery at a speed of 15 knots. It is assumed here that is 15 knots relative to their scale; otherwise we would not see the long scenes in the movie showing *Proteus* navigating in the artery.

Fifteen knots (knots are used in the navy to indicate the speed of the vessel) corresponds to 15 nautical miles per hour. One nautical mile or one minute of latitude (referring to the circumference of the Earth), which is a bit longer than the terrestrial mile, corresponds to 1852 meters. Knowing that the crew members have an initial average height (before being miniaturized) around 175 cm, we could assume that the length of *Proteus* was around five times that especially if you look at the scene 00:19:04 showing the length of *Proteus* with respect to the height of a human adult. That would mean that the length of *Proteus* was initially around 8.75 meters. So 15 knots would mean that *Proteus* was in fact traveling at 7.72 body lengths per second when in the carotid artery. Compared to our first submarine that travel in the carotid artery at 66 body lengths per second, we can say that the reality has surpassed fiction, at least for the speed of the submarine.

As another note, if the average size of the crew members would be equivalent to the average size of our crew members being 1.5 micrometers, then the length of *Proteus* would be five times that, i.e., 7.5 micrometers, about the size of the submarine depicted in Figure 7.1. Since in the scene with regard to the size of the red blood cells, *Proteus* appears to be longer than that, this suggests that our crew members are even smaller than the miniaturized crew members in the movie. This may also be an illusion in the scene with the red blood cells not being sufficient close to the submarine. But in all cases, it appears that *Proteus* might have a minimum length equivalent to 2 to 3 red blood cells to potentially up to approximately 30 micrometers as mentioned earlier.

Another issue with the velocity of our submarine is concerned with the possibility of the submarine to aggregate with other submarines, if more than one submarine is injected at the same time. This is illustrated schematically in Fig. 7.2.

In Fig. 7.2, if the velocity of the submarines is not sufficient, the dipolar interactions between submarines due to the dipole field around

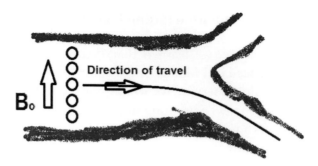

Figure 7.2 Simple schematic representing the formation of an aggregate of microscopic submarines due to dipolar interactions.

each submarine could ease the formation of chain-like aggregates with the easy (longitudinal) axis parallel to the B_0 field. If such an interactive force between the submarines is too strong, then such a formation could be jammed at the next bifurcation, depending on the direction of travel with regard to the direction of the uniform field inside the tunnel of the MRI scanner. Increasing the speed could break such an aggregate.

On the other hand, such an aggregation will increase the effective volume of magnetic material or nanoengines working together, which may help to steer the submarines in the right direction. But if this is done, the submarines will typically be synthesized such that the interactive dipolar force produced by each submarine will be at the level required to maintain such dipolar interactive force within an acceptable range. Such acceptable range corresponds to dipolar interactive forces that are strong enough to maintain the microscopic submarines in formation (similar to formations used by military ships) during travel while being sufficiently weak to break the formation (aggregation) at each vessel bifurcation in order to avoid an unplanned embolization.

7.4 Distance to the Next Branching Artery (Scene 00:41:32–00:41:35)

In this scene, it is reported that *Proteus* is at 2 minutes at the present estimated speed of 7.72 body lengths per second. Assuming that *Proteus* is 7.5 micrometers long, the distance to the next bifurcation would be just below 7 mm, which is possible. The one aspect that is highly coherent with our approach is that the injection was done relatively far for the next bifurcation in the artery. Indeed, we always attend to inject our microscopic submarines as far as possible to the next bifurcation to allow the maximum distance and time to steer the submarines. Doing this allows us to maximize the likelihood of steering the submarines to the appropriate targeted branch by providing sufficient time for steering. In the navy, we refer to this as the ship's turning data which are taken into account when navigating the ship in tight navigable channels.

7.5 Submarine not Responding Due to a Strong Blood Flow (Scene 00:41:35–00:43:25)

At the end of this scene, we learn that *Proteus* is not responding when attempting to steer it due to a too high blood flow. This is also typically the case for our submarines when navigating in the artery. Assuming that the blood flow can reach 0.3 meter per second in the artery, for our largest submarines of 200 micrometers, this means that our microscopic submarine would travel at a velocity corresponding to 1500 body lengths per second, way over the maximum velocity of *Proteus* estimated at around 8 body lengths per second. There are two approaches that can be used in such a case. The first one is increasing the steering force of our submarine during navigation; this can be done by a technique that we put in place and is known as dipole field navigation (DFN), which will be explained later in this book. For the standard navigation approach known as magnetic resonance navigation (MRN), a technique also pioneered by our group, maintaining the steering response can be achieved by reducing the flow using, for instance, the balloon catheter shown in Fig. 6.3, which is also used to inject the submarines. The last part of the scene really shows how fast the flow can be if it is not controlled.

Just after this scene, we see *Proteus* transiting through a hole in the arterial wall. As a note, it was possible only because the hole was already there. In our cases and probably also in the movie, our submarines at such a small scale do not have sufficient power to perforate the arterial walls.

It should also be noted that turbulent flows as seen in this scene only occur in fluid conditions where the Reynolds number is typically greater than 4000. The Reynolds number is defined as the ratio of inertial forces to viscous forces. For example, the Reynolds number of a man (not miniaturized) swimming in water is 10,000, whereas if the same swimmer would be reduced to the size of bacteria, the Reynolds number of the miniaturized swimmer with the same swimming performance as bacteria in water would be 0.0001. At such a scale, water or blood will feel extremely viscous, like a very thick liquid, making swimming for such miniaturized divers extremely difficult. The good news is that below 2300, the flow will be laminar and very smooth without any turbulences. The region between 2300 and 4000 is known as the transient regime. It should also be noted that turbulences that only occur in larger blood vessels can help to break or prevent the aggregations of submarines such as in arterial bifurcations as depicted in the simple diagram in Fig. 7.2.

8

Tracking the Position of the Submarine

W HILE THE SUBMARINE *Proteus* navigates, its position is being tracked during its journey in the human body. As in the movie, being able to track the position of our microscopic submarines also proves to be essential. In this chapter you will find how tracking the position of such miniature submarines is really being done.

8.1 Position of the Submarine Being Indicated on a Large Map (Scene 00:45:30–00:45:58)

This is a scene that we saw several times so far in the movie, where a man is moving some kind of large marker to indicate the position of the submarine *Proteus* on a very large map of the vascular network. The use of such a large map is probably done in the movie so that others can see it, since plotting the position on a small display would be hard for everyone to see. In the movie, this seems to be a good idea when we look at the relatively large number of persons being involved.

But the number of persons involved in our setup during an intervention is much less than in the movie, and therefore, plotting the position of our microscopic submarines on a large map for this

A Microscopic Submarine in My Blood: Science Based on Fantastic Voyage
Sylvain Martel
Copyright © 2016 Pan Stanford Publishing Pte. Ltd.
ISBN 978-981-4745-78-9 (Hardcover), 978-981-4745-79-6 (eBook)
www.panstanford.com

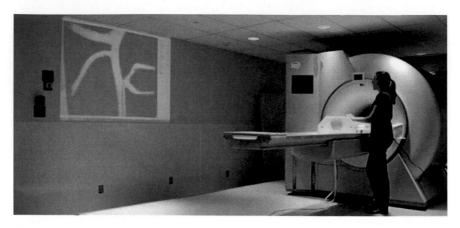

Figure 8.1 Photograph showing an MRI-tracked position plotted on a computer-generated map of an arterial network being projected at a larger scale on one of the walls of the interventional room.

purpose is not really a requirement. But we are still, in some cases, projecting the tracked position on a large map. One example is depicted in Fig. 8.1, where a computer-generated display of the tracked position and the map (vascular network) are being projected on a wall in our interventional room. In the photograph, we see someone next to our previous 1.5 T clinical MRI scanner (that is replaced in the new setup shown in Fig. 2.1 by a 3 T clinical MRI scanner) inserting a guidewire into an artificial arterial network known as a phantom. In this particular photograph, a phantom is used by a biomedical engineering PhD student to test a particular navigation and tracking sequence before being applied on a live animal. As she is inserting the guidewire, she is looking at the projected position of the guidewire gathered by the MRI scanner. This projected image provides the feedback information necessary to guide her during the procedure.

Although x-rays are typically used to track the insertion of a guidewire in an artery, MRI was used, instead, in the example depicted in Fig. 8.1, because unlike typical guidewires, the guidewires used by our group can be magnetically steered along the planned path in the arteries. One example of such a magnetically steerable guidewire is depicted in Fig. 8.2.

Typically, when you introduce a guidewire in an artery, you push the guidewire further along the blood vessel. But when the guidewire

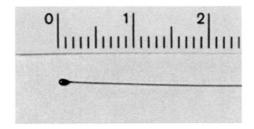

Figure 8.2 Photograph showing an example of the tip portion of a guidewire being inserted with the assistance of the MRI scanner. The magnetic tip of the guidewire in this particular example (at the zero mark at the end of the wire) has a diameter of 900 micrometers (0.9 mm). From "In vivo demonstration of magnetic guidewire steerability in a MRI system with additional gradient coils", *Medical Physics* 42:969-976, 2015, with permission from American Association of Physicists in Medicine.

is made thinner to allow it to be introduced in narrower blood vessels, the stiffness of the guidewire will decrease significantly. As you go deeper, the friction with the vessels' walls opposes the displacement of the guidewire and the latter begins to bend, preventing it from going further. Making it stiffer would increase the risk of perforation in the arterial walls, while reducing its flexibility. This will, in turn, prevent it from taking turns when going deeper into the arteries.

One strategy used here to prevent this from occurring, while allowing a thinner guidewire to navigate deeper into the vascular network, is to apply a directional magnetic pulling force at the tip of the guidewire. Such a magnetic pulling force is provided here by the MRI scanner. The advantage of a large computer-generated image being projected on the wall instead of the gigantic and static map depicted in the scene of the movie is that it can be dynamically changed to provide another image that will provide suitable information at every step of the operation. One example is shown in Fig. 8.3, where the projected image has been used to ease the task of the surgeon by showing only the information required to perform the operation.

But we found that it does not matter how good the computer-generated images are; the reality is that doctors always want to see the "real thing" just to make sure. Therefore, Fig. 8.4 depicts an example of that where the real image is also projected next to the computer-generated images after the targeted location has been reached. You can see the MRI artifact created by the magnetic tip of the guidewire. Such an artifact appearing like a black spot is shown

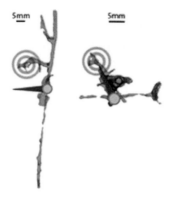

Figure 8.3 Computer-generated image projected on the wall of the position of the magnetic tip of a guidewire in an artery. Here the position of the magnetic tip of the guidewire is shown as a green dot. The arrow connected to the green dot indicates the direction of the applied magnetic field to steer or pull the tip of the guidewire in the right direction within the arterial network. The targeted location where the tip of the guidewire should be located is indicated here by a symbol consisting of two circles. Although the arteries could be represented as in Fig. 8.1, here they are represented as the result of the image reconstruction process taken from the MRI of the blood vessels. From "In vivo demonstration of magnetic guidewire steerability in a MRI system with additional gradient coils", *Medical Physics* 42:969-976, 2015, with permission from American Association of Physicists in Medicine.

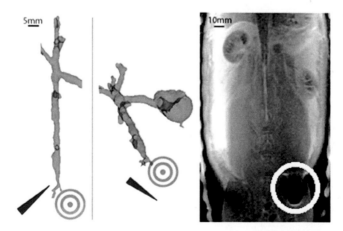

Figure 8.4 Real acquired image put next to the computer-generated images to help confirm the tracking position at the targeted physiological location. The real image is used to visualize the physiological features and the image artifact created by the magnetic object (shown in the white circle). Adapted from "In vivo demonstration of magnetic guidewire steerability in a MRI system with additional gradient coils", *Medical Physics* 42:969-976, 2015, with permission from American Association of Physicists in Medicine.

in a white circle that has been intentionally added to help the reader to identify it on the image.

A guidewire is typically introduced first to later guide a catheter or microcatheter in the arterial network. Some of you might ask at this point the reason behind the use a magnetic guidewire instead of introducing first a magnetic catheter being magnetically guided the same way as the guidewire. The reason is simple. The catheter is used (when the use of a syringe is not an option) in our particular applications to inject the microscopic submarines into the arterial network, as discussed in Section 6.1.2. If the tip of the catheter would be magnetized, it would attract the magnetic submarines to the magnetized tip, preventing them from being injected in the arterial network. We introduce, instead, a magnetic guidewire. When the tip of the guidewire reaches the targeted location, as in the example depicted in Fig. 8.4, the catheter is introduced with the guidewire entering the inner canal of the catheter (or microcatheter). Then the catheter is pushed along the guidewire until it reaches the targeted location. When this occurs, the guidewire is retrieved and the team is then ready to initiate the injection phase. Because of the larger diameter of the microcatheter and the friction force with the vessel walls (especially in tortuous vessels) preventing it from going deeper into narrower vessels, the injection of magnetic microscopic submarines becomes essential if deeper regions in the vascular network must be targeted.

The same type of large display can also be used to track the position of our microscopic submarines. This would be especially interesting for the medical staff as they could confirm that the submarines reached their final targeted destinations. In such a case, large images similar to the ones depicted in Fig. 8.4 could be projected to allow the doctors to confirm that, indeed, the targeted location has been reached. In this case, the position of the submarines in the MRI-acquired image would appear like the image artifact shown in the white circle but smaller.

Going back to the scene in the movie, you can observe that although the use of a larger display is in concordance with our practices, it is not accessible in the movie (as it is in our case) to the medical specialists who are still in another room, as shown in the beginning of the next scene.

8.2   The Tracking System (Scene 00:46:00–00:46:36)

In this particular scene, the only thing that is coherent with us is the hand holding a cup of coffee. Indeed, we drink a lot of coffee. Personally I do not put sugar in coffee, as in the movie; I like it black.

Throughout the scene, you can see an assembly of several moving small-radar antennas around the patient's head which are used to track the position of the submarine *Proteus*. The movie claims that this system can detect the radioactivity generated by the microscopic particle used to power the submarine *Proteus*.

8.2.1   *Tracking with the Technology as in the Movie (PET-Based Tracking)*

The tracking approach proposed in the movie is indeed possible, and it is known today as positron emission tomography (PET). To do such a PET-based tracking, the specialists in the movie would need a PET scanner. The first PET scanner was available in 1975, only nine years after what happened in the scene of the movie. This explains, in part, why the vision of Hollywood showing the small radars around the head of the patient was a bit offset, and I say "a bit" because I do not want to hit too hard on those visionaries. Figure 8.5 shows a photograph of one of the first PET scanners. As you can see, it is very different and much bulkier than the set of small radars that we see in the movie. In fact, today clinical PET scanners still look very similar to the one depicted in Fig. 8.5, and indeed, the system looks much more different than in the movie.

Furthermore, the first PET scanners had a very small number of radiation sensors, limiting the quality of the images acquired. Considering the relatively low numbers of small radars used as radiation sensors (receivers) in the movie, we could expect, if in fact these radars could do what Hollywood claims, that tracking with sufficient resolution would be a real challenge.

Indeed, the resolution of modern PET scanners is much lower than the resolution of other medical imaging modalities, including MRI. They are also must slower, so the frequency of the blip that we often

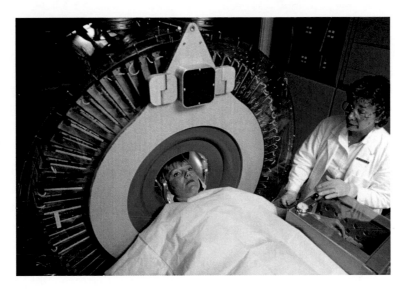

Figure 8.5 Photograph of a PET scanner positioned around a patient's head as the position of the tracking system depicted in the movie. Image taken from http://www.cerebromente.org.br/n01/pet/pet_hist.htm (copyright © 2012, with permission from Science Photo Library Ltd.).

see on the TV monitor in the movie to indicate the update position of *Proteus* would be much slower.

Furthermore, since we already use an MRI scanner to power our microscopic submarines, it made more sense for us to use MRI-based tracking instead of PET-based tracking, as in the movie. But since PET-MRI scanners are now available, combining both PET and MRI, MRI- and PET-based tracking remains a possibility in our case by making our microscopic submarines both superparamagnetic and radioactive. But considering that MRI has many advantages compared to PET (such as better spatial resolution; faster speed to acquire images or track information; the fact that it is the best medical imaging modality to image human soft tissues, which is critical in our case; and the fact that it is extremely sensitive to magnetic particles, making it highly effective at detecting our microscopic submarines, all this without the need for radioactive materials), it makes you question the need to add PET capability for tracking purposes. This is one reason that a PET scanner is not presently planned in the intervention room (Fig. 2.2).

8.2.2   *MRI vs. Ultrasound, X-Ray, and OCT-Based Tracking*

From our experience, we found that not only did MRI proves to be the best platform for navigating microscopic magnetic submarines in the bloodstream, it also provides the best medical imaging modality to track and detect the position of these microscopic submarines. Ultrasound systems are definitively not a good option since they have the worst resolution, making them unsuitable to detect such small submarines. X-rays, including computed tomography (CT) scanners, have a resolution at the human scale of approximately 200 micrometers, just at the limit to theoretically detect our larger version of navigable submarines. But this is in theory, and practically, this may prove to be very difficult. Furthermore, it limits the level of miniaturization of our submarines. Optical coherence tomography (OCT) has a very good resolution and can detect very small submarines, but unlike the other imaging or detection modalities, including MRI, it cannot penetrate very deep into the body. This is typical of all light-based detection systems. Although the one with a wavelength in the infrared spectrum can penetrate deeper into physiological tissues, the maximum penetration depth is still insufficient for this particular application. This leaves MRI as the best tracking modality for our magnetic submarines. Such a conclusion is further explained in the next subsection.

8.2.3   *Fundamental Principle of Detecting the Location of a Microscopic Submarine in the Body*

Since the detection and tracking of a microscopic submarine are presently based on MRI technology, before going further let me first explain how MRI works. Since many books already explain how MRI works in details, I am just briefly describing the basic principle here and how it is applicable to the localization of our submarines in the body.

The patient is first placed in the tunnel of a clinical MRI scanner (such as the one in the interventional room in Fig. 2.2), where there is a high magnitude (typically 1.5 T or 3 T) static (uniform) field. This field, as I mentioned earlier, is known as the $\mathbf{B}_0$ field, and it is parallel to what is referred to as the z axis, which goes horizontally from the entrance of the tunnel to its far end. As much as 60% of our body is constituted of water, and the water molecule is known to be made of one atom of oxygen with two atoms of hydrogen. So, this is to say that the human body has a huge amount of hydrogen atoms. Each hydrogen atom has one proton, which is a positively electrically charged subatomic particle that can be influenced by an electric or a magnetic field.

Due to the direction of the $\mathbf{B}_0$ magnetic field along the z axis influencing the movement of the protons, the hydrogen atoms will be aligned in the direction of the z axis, which is parallel to the static magnetic field $\mathbf{B}_0$ of the MRI scanner. Now imagine that you "push" the protons of the hydrogen atoms in order to align them in a different direction (e.g., 90° or 180° from the original direction). This will require us to apply some form of directional energy capable of doing just that. This is actually done by transmitting a high-powered radio frequency (RF) pulse. To avoid interference with other electronic equipment located in adjacent rooms, our entire interventional room, as shown in Fig. 2.2, is shielded using copper plates inside the walls and a grid or mesh in the large window between the interventional room and the control room. The grid in the window is based on the same principle as the grid or mesh placed in the window of a microwave oven. Interesting also, when you look at the short scene of the movie starting at 00:25:15, is the fact that a grid is also used but it is not clear for which purpose. If it would be used to shield, the wavelength of the signal would have a very long wavelength, considering the large gap between the meshes.

Immediately after the transmission of the RF pulse, the alignment of the hydrogen atoms would have flipped to the new direction. At this exact instant, the hydrogen atoms begin to orient back along the z axis since the hydrogen protons will come back to their original position, being forced by the high-magnitude directional $\mathbf{B}_0$ field of the MRI scanner. When the hydrogen atoms flip back to their original orientation parallel to the z axis, they release the stored energy that was used to flip them initially. This energy being released is captured by special antenna and amplified to be processed by a computer program. The time it takes to come back to the original orientation is known as the relaxation time. This relaxation time depends on the magnitude of the B_0 field. Indeed, a stronger field will impose a stronger force to reorient the atoms, and therefore, the relaxation time will be shorter. The relaxation time will also depend on the type of atoms.

Inside a magnetic object such as the magnetic particles embedded in a microscopic submarine operating inside a patient placed in the tunnel of an MRI scanner, the magnetic field inside the magnetic particles used to propel the submarine will be stronger than the surrounding magnetic field. So when the RF pulse is transmitted, the atomic structures in these particles remain oriented towards the z axis, and because there is no flipping and therefore no relaxation, there is no energy being transmitted back to the receiving antenna. In other words, there will be a loss of signal at the location of the submarine.

But here is what makes more interesting the use of MRI to detect such microscopic submarines. The net magnetic dipole field generated around the submarine by the magnetic nanoparticles being at saturation magnetization is much larger than the size of the submarine itself and it is stronger than the $\mathbf{B}_0$ field of the MRI scanner. So when the RF pulse is applied, this strong magnetic dipole field oriented towards the z axis maintains the hydrogen atoms aligned (or approximately aligned if further away from the submarine) along the z axis, resulting in a loss a signal around the submarine as well. All this volume where there is a loss of signal appears as a black image artifact, such as the one depicted in Fig. 8.4, that is much larger than the microscopic submarine itself. From our experience, such an image artifact could be 50 to 70 times the overall size of one of our microscopic submarines when placed in a 1.5 T MRI scanner, for instance.

If we consider that the spatial resolution or the size of a voxel (a 3D pixel) for a 3 T clinical MRI scanner is approximately 500 micrometers, this means that a microscopic submarine as small as 10 micrometers (roughly one-tenth the thickness of a human hair) could theoretically be detected by MRI. This is much superior compared to a x-ray or a CT scanner when we consider that the spatial resolution of an x-ray or a CT scanner is roughly 200 micrometers, which means that any submarines with an overall size of less than 200 micrometers cannot be detected.

Figure 8.6 shows some examples demonstrating the capability of a 1.5 T or a 3 T clinical MRI scanner to detect miniature magnetic submarines. As you can see, both can detect a submarine as small as

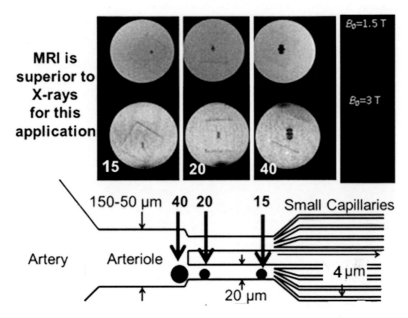

Figure 8.6 MRI-based detection using a 1.5 T or a 3 T clinical scanner of a single magnetic submarine with an overall size ranging from 40 down to 15 micrometers.

15 micrometers, a size sufficiently small to detect a single submarine located deeply in the vascular network down to the large capillaries but out of reach of the small capillaries closer to a tumor, for instance. This is where our crew members (each with an overall size between 1 and 2 micrometers) will swim after leaving the submarine.

Another example is depicted in Fig. 8.7, where a few of the 50-micrometers-in-diameter magnetic submarines (the same as in Fig. 4.1) are detected and localized in a targeted lobe in the liver, where they deliver therapeutics after being navigated in the hepatic artery. The image taken with a 1.5 T clinical MRI scanner really shows that submarines as small as half the diameter of a human hair can be detected and their position determined in deep regions of the body.

Now that we know that the microscopic submarines can be detected anywhere in the body, the remaining question is, how do we calculate the exact location of these submarines in the body so that such a position can be plotted on the computer displays or used by the navigation computer to calculate the next action to be taken to

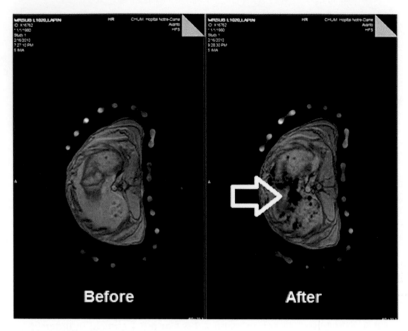

Figure 8.7 MRI of the liver taken before and after a few microscopic submarines had reached the targeted region, following navigation in the hepatic artery.

maintain the submarines on the planned trajectory? This is done by the superposition of a magnetic gradient on top of the uniform $\mathbf{B}_0$ field. An MRI scanner can produce such a magnetic field that varies in magnitude over distance (magnetic gradient) using electromagnetic coils in the core (periphery surrounding the tunnel) of the MRI scanner. These electromagnetic coils are not superconductors, because the gradient field must be modulated at relatively high frequencies, which is not possible using superconductive technology. As such, the gradient field is limited to approximately 0.04 Tesla per meter in a standard clinical MRI scanner. Such a value determines the thickness of the image slice, which determines the spatial resolution of the scanner. A higher gradient means better spatial resolution. An MRI scanner can generate gradients in any 3D direction. By applying a gradient on the x, y, and z axes, which adds to the strength of the uniform field, the magnitude of the magnetic field will vary at some locations in the body of the patient, which will have an impact on the relaxation time. Knowing the expected relaxation time in such specific 3D slices, the image can be reconstructed using the appropriate mathematical equations. In turn, knowing the ratio of electrical currents circulating in the x, y, and z gradient coils, we can determine mathematically the exact position within the spatial resolution of the MRI system of our submarine in 3D space.

8.3   The Tracking Position of the Submarine Shown on the Display (Scene 00:46:47–00:47:00)

This is one of several scenes where you see the position of the submarine *Proteus* being represented by a flashing blip superimposed on a map of the blood vessels appearing on a TV monitor. In Fig. 8.8, you see

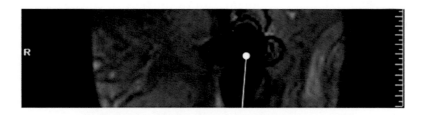

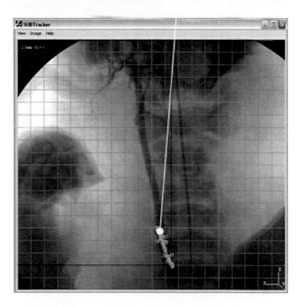

Figure 8.8 Image of the first submarine being tracked while it is navigating in the blood vessel of a living animal.

one example of how the position of our submarine can be shown on a computer display.

In this example, these images were taken when we performed the first demonstration of the navigation of an untethered object in the blood vessel of a living animal in November 2006. Here, the 1.5 mm magnetic submarine was travelling in the carotid artery at a velocity of 10 cm per second (over 66 body lengths per second). The large MRI artifact depicted on the top image of Fig. 8.8 covers the surrounding image of the physiological features, making it hard to someone to really see where the submarine is in the body. So, a special algorithm was developed and used to determine the center of gravity of this artifact. Then, as depicted in the lower image in Fig. 8.8, such center of gravity was indicated as a computer-generated symbol superimposed on a preacquired image of the interior of the body made by combining x-ray and MRI. Typically only MRI is used since information about the location of the bones such as the spinal cord shown in the image is not necessary in most cases. The image at the bottom of the Fig. 8.8 depicts the position of the submarine traveling along the carotid artery, being updated 20 to 30 times per second. But with a much smaller submarine, the signal captured by the MRI scanner would be much weaker, which would require more time for detecting and localizing the position of the submarine. In turn this would increase the time between two consecutive position updates on the display (the time between two successive blips in the movie).

9

Propulsion and Steering

I N THIS CHAPTER, we are taking a better look at the propulsion and steering of the submarine *Proteus* and comparing it to how it is done with our microscopic submarines.

9.1 Go with the Flow (Scene 00:47:27–00:47:30)

In this very short scene of the movie, the pilot of *Proteus* is instructed to maneuver the submarine in the direction of the blood flow and to drift with it. This is exactly how we do it. Indeed, the blood flow constitutes a huge power source that is exploited to propel at a very fast velocity our microscopic submarines. Indeed, the heart is a gigantic and powerful pumping machine compared to the size of our submarines. Not exploiting such an available propelling force would be a real waste. But to take advantage of this available propelling force means that we should always plan navigation according to the direction of the blood flow, and in reality, we do not really have the choice. When in the scene at 00:45:35 it is said, "We can't go back," well, we can't go back either.

Indeed, the simple equation depicted in Fig. 9.1 is typically used to calculate the total propelling force that the superparamagnetic

A Microscopic Submarine in My Blood: Science Based on Fantastic Voyage
Sylvain Martel
Copyright © 2016 Pan Stanford Publishing Pte. Ltd.
ISBN 978-981-4745-78-9 (Hardcover), 978-981-4745-79-6 (eBook)
www.panstanford.com

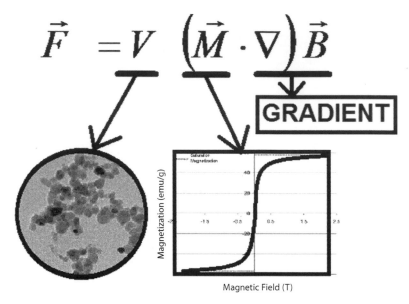

Figure 9.1 Fundamental equation to calculate the propelling/steering force of a microscopic submarine relying on superparamagnetic nanoparticles.

nanoparticles acting as nanopropulsion engines can provide for each of our magnetic microscopic submarines. Not to worry, this is the only equation that is described in this book, but I believe that it is simple enough. I put it there because it explains very well how the propulsive force can be generated.

The equation in Fig. 9.1 tells you that the propulsive force of our microscopic submarine depends on three factors or variables. The first one is the total volume (V) of the accumulated superparamagnetic nanoengines of our submarine. In other words, it is the total volume of all superparamagnetic nanoparticles embedded in each submarine. So the higher the value of V, the more the propulsive force that can be produced from each microscopic submarine. This is obvious when you consider that bigger engines typically produce more force than a smaller engine, while the total propelling force can also be increased with a sufficient number of smaller engines working in unison (superparamagnetic nanoparticles) instead of one large engine (ferromagnetic particle). But because V scales down very fast (at a cubic rate and not linearly), the force decreases extremely fast with miniaturization. It then becomes difficult to maintain a sufficiently high propulsive force as the overall size of the submarine reduces. At some point, V becomes so small that achieving sufficient propelling forces to navigate effectively in the vascular network becomes almost impossible. But we do not really have the choice since the submarine must be sufficiently small to transit through narrowed blood vessels. With experience, we found that the minimum size that the submarine can be to effectively navigate in the arteries with half its volume used for iron oxide superparamagnetic nanopropelling engines would be down to a few tens of micrometers in diameter. This is the main reason that when the submarine becomes smaller, we use the blood flow to propel the submarine and to steer the submarine along the planned trajectory. Indeed, the 1.5-mm-diameter chrome steel submarine first used to navigate in the carotid artery had a sufficiently large V to go back and forth at a high velocity along the artery. But for the much smaller submarines depicted in Fig. 4.1, for example, the blood flow was used for propulsion, while the magnetic force available was used for steering. An example of the submarines being propelled by the blood flow and steered by the force provided by the onboard nanoparticles is depicted in Fig. 9.2. As depicted in the MRI images, steering ensures that the submarines take the appropriate branching vessel. The arrow on top of a variable in the above equation means that this variable is a vector that points to a given direction.

The next variable that influences the force is the magnetization of the superparamagnetic nanoparticles or nanoengines embedded in each submarine. It is represented in the equation as M. The magnetization depends on the type of superparamagnetic magnetic material (e.g., iron oxide versus iron cobalt), how it was synthesized, and the magnetic field strength that the submarine in being exposed to, where ideally such field strength should be sufficiently high to bring M at saturation magnetization. As mentioned earlier, the latter is achieved here by the $\mathbf{B}_0$ field of the clinical MRI scanner. Therefore, improvement for M above what is possible today could only be done with the discovery or synthesis of a new material or alloy having a higher saturation magnetization level.

The remaining variable is the magnetic gradient, or variation of the magnetic field strength over distance. In the equation depicted

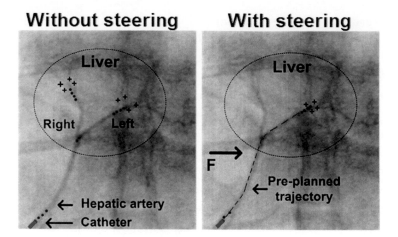

Figure 9.2 Targeting a specific region in the liver with and without steering forces on the submarines depicted in Fig. 4.1 during their transit in the hepatic artery at 10 cm below the skin. From *Biomaterials*, 32(13):3481–3486, copyright © 2011, with permission from Elsevier.

in Fig. 9.1, it is the symbol that looks like a reverse triangle with $\vec{B}$. Unfortunately, since the magnetic field strength is known to decay rapidly with distance from the magnetization source, it proves hard to achieve very high gradients over a distance sufficient to cover a human adult. Different approaches to generate such magnetic gradients will be discussed in the following sections, each with their advantages and disadvantages.

But by combining all these variables together, we concluded that for small magnetic submarines intended to navigate in narrower blood vessels, the propelling forces was really insufficient, and as such, we also, as in the movie, had to go with the flow. But to reach our targets this is typically acceptable. Indeed, it is the same blood flow that brings the nutrients and oxygen to the tumor. Therefore, by following the blood flow while taking the right-branching blood vessels, these microscopic submarines can come sufficiently close to the tumoral mass to release our microscopic crew members at swimming distance from the target.

9.2 Pulsatile Flow (Scene 00:49:37–00:50:04)

In this scene of the movie, the crew members of *Proteus* are feeling the pulsatile flow created by the heart as they approach the right atrium.

Again, this may sound a bit strange, but we also, in some particular cases, intentionally create a pulsatile flow similar to the one in the movie but independent of the one created by the heart.

This is particularly true when the buoyancy of the microscopic submarines is reduced due to an additional weight and cannot be compensated (due to a limitation of the overall dimension of the submarines) by techniques such as the addition of a buoyant layer (mimicking the ballast in real submarines). This extra weight can be caused by additional superparamagnetic nanoengines (compared to other versions) required to increase the propulsion or steering power of each submarine. In this case, especially when the velocity is reduced, such submarines will go down until they are deposited at the bottom of the artery. While touching the bottom of the artery, the friction force with the vessel wall, combined with the lower velocity of the blood flow near the wall due to a parabolic flow distribution in the arteries where the flow is maximal along the center of the artery, can prevent the submarines from proceeding further. By modulating the blood flow to create a pulsatile flow similar to the one in the movie using a computer-controlled balloon catheter (Fig. 6.3) or through a special vibration system, such a potential problem can be avoided.

9.3 Propelling and Steering Systems
(Scene 00:50:13–00:50:16)

In this very short scene, we see clearly one of the two propulsion engines of *Proteus* expelling a jet stream as the submarine accelerates to reach maximum velocity. It is really good that the scenario did not put propellers instead of these microjets. Indeed, because of the Reynolds number introduced earlier in Section 7.5, which would be very low in this situation, propellers similar to the ones used to propel ships would not work. In this respect, the use of jets is a very good choice.

9.3.1 *Microjets*

But more interesting is the fact that jet-propelled miniature submarines designed to operate in the body already exist. Figure 9.3 depicts a photograph of a miniature microjet engine.

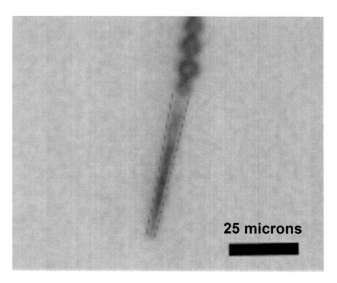

Figure 9.3 Close-up view of a miniature microjet engine in action; the exhausted jet consists of microbubbles on the back of the engine, moving in the reciprocal direction at 100 micrometers (microns) per second. From *IEEE Trans. Robotics*, 30(1):40–48, copyright © 2014, with permission from IEEE.

The miniature microjet engine similar to the one depicted in the scene of the movie showing one of the two jet engines of *Proteus* can be directionally controlled (steered) by making its structure magnetic. By applying a relatively weak directional magnetic field, a directional torque would be induced on the microjet, forcing the latter to be propelled in the direction of the applied magnetic field. Besides the fact that the propulsion rate cannot be controlled during navigation, the problem with a microjet is that its operation depends on the characteristics of the surrounding fluid. For instance, for the version depicted in Fig. 9.3 to operate, the aqueous solution must contain 1% hydrogen peroxide (H_2O_2) and 1% sodium dodecyl sulfate (SDS), which are not present in human blood.

9.3.2 Magnetic Resonance Navigation

The main advantages of using a magnetic field, instead, are that magnetic actuation (propulsion and steering) is independent of the characteristics of the fluid medium (blood, plasma, etc.) and the magnetic field is not affected by physiological tissue density. The main method being used by our group to propel and/or steer (actuate) microscopic superparamagnetic submarines for navigation purpose in the bloodstream is a technique pioneered by our group, known as magnetic resonance navigation (MRN).

MRN exploits the magnetic environment inside the tunnel of a clinical MRI scanner to actuate these microscopic submarines for navigating them along a planned path in the blood vessels. As mentioned in Chapter 4, the uniform B_0 field is used to bring the superparamagnetic nanoparticles acting as propelling nanoengines at saturation magnetization (by analogy to engines turning at a full revolution per minute [RPM] but with the clutch or transmission in the neutral position). To move the submarine in a given direction, a directional magnetic force must be generated. To generate such a directional force, we need the addition of a directional magnetic gradient, as shown in the equation in Fig. 9.1. In MRN, this gradient is generated by the MRI gradient coils, as depicted in Fig. 9.4 (the ones used to select an image slice, as mentioned in Section 8.2.3). Since 3D image slices can be produced in MRI, these electromagnetic coils can also be used to move the microscopic submarines in any directions in 3D space. Increasing or decreasing the current in the gradient coils is like pressing or depressing the gas pedal to accelerate or decelerate your car. To turn the steering wheel, you will have to modify the ratio of electrical currents circulating in the x, y, and z gradient imaging coils of the MRI scanner.

With MRN using a conventional MRI scanner, you have the nanoengines in your submarine running at full regime with a typical magnetic field strength of 3 T, but the magnetic gradient available is relatively low, typically at 0.04 T/m. In the past, we have increased the gradient up to 0.4 T/m (10 times the typical value in clinical MRI scanners) with a special gradient insert, but the problem, as depicted in Fig. 9.5, is that the inner diameter of the tunnel could be made sufficient for the head but could not accommodate the torso of a human adult. The gradient insert depicted in Fig. 9.5 consists of two type of coils, namely a steering gradient coil (SGC) assembly and an imaging gradient coil (IGC) assembly. The advantage of the SGC is that it can maintain

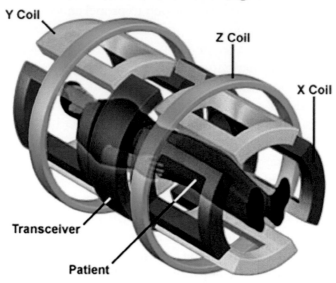

Figure 9.4 Gradient coils in a clinical MRI scanner that can be used not only for imaging but also to move microscopic submarines in any direction. Image taken from magnet.fsu.edu (with permission from the National High Magnetic Field Laboratory).

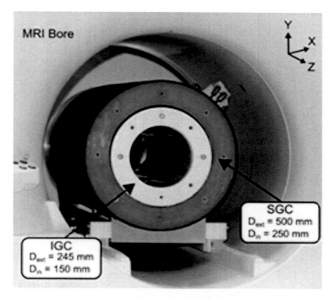

Figure 9.5 A gradient insert can be used to increase the gradient of a clinical MRI scanner but at the price of a decrease of the inner diameter of the tunnel.

a gradient for a long time without generating excessive heat. But the SGC cannot change the direction of the microscopic submarines very fast. The IGC can change the direction of the microscopic submarines at a very fast rate, but it cannot maintain the directional gradient very long. As such a series of pulsed directional gradients is generated instead, but by doing so, the IGC tends to overheat if such sequences of pulses are maintained for approximately 2 minutes, at which time steering the submarine cannot be done until the coils cool down, which will take approximately 20 minutes. These numbers depend on the characteristics of the coils and may vary between different coil assemblies. The same overheating problem is present when using the imaging gradient coils of the MRI scanner for propelling or steering our microscopic submarines for a relatively long time.

In Fig. 9.5, the IGC is inside the SGC since the latter would prevent imaging to be taken if located inside the IGC. The SGC cannot be used for imaging or tracking the microscopic submarines, and as such, it is often removed to allow the implementation of an ICG assembly with an inner diameter that would be sufficient to insert the head of a patient.

9.3.3 *Dipole Field Navigation*

Dipole field navigation (DFN) is a more recent approach pioneered by us. The basic idea is to distort the uniform $\mathbf{B}_0$ field inside the tunnel of the clinical MRI scanner in such a way that it will produce some kind of magnetic path along the blood vessels that our microscopic submarines will be forced to follow. Such distortion is done by placing relatively large (up to a few centimeters in diameter) ferromagnetic spheres at specific locations inside the tunnel of the MRI scanner. The advantage of this approach is that the distortion created by the magnetic dipole of each sphere can produce a magnetic gradient deep in the human body, about 10 times the strength of the gradients that can be generated by the MRI coils of the scanner. The other advantage is that by doing so, DFN maintains the high field strength of the $\mathbf{B}_0$ field to bring the superparamagnetic nanoparticles acting as nanoengines for our microscopic submarines at saturation magnetization (full regime) (Fig. 9.6).

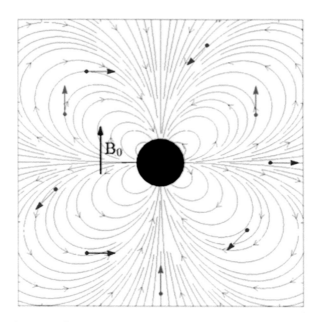

Figure 9.6 Image of the dipole magnetic field around a relatively large ferromagnetic sphere placed in the tunnel of an MRI scanner; such a dipole distorts the strong homogeneous field of the scanner, which, in turn, translates into high directional gradients (shown by the colored small arrows) that can be used to propel or steer the microscopic submarines in the bloodstream. The B_0 field strength required to bring superparamagnetic nanoparticles at saturation magnetization is also maintained. From *IEEE Trans. Robotics*, 31(6):1353–1363, copyright © 2015, with permission from IEEE.

The inner diameter of the tunnel can also remain sufficiently large to accommodate a human adult if a large-bore MRI scanner is used to allow sufficient space to put an insert containing the ferromagnetic spheres. This explains, in part, the reason that the MRI scanner depicted in Fig. 2.1 is, in fact, a large-bore 3 T clinical MRI scanner.

With a proper placement of the ferromagnetic spheres in the insert surrounding the patient, the resulting directional gradients will steer the microscopic submarines along the preplanned trajectory in the arterial network, as depicted in Fig. 9.7.

Prior to beginning DFN, the exact placements of the large ferromagnetic spheres are determined mathematically using special

Figure 9.7 Example of microscopic submarines being steering along the planned path using DFN.

software programs specially developed for this purpose. The computer program will calculate such optimal placements on the basis of the planned trajectory being plotted on an image acquired of the arterial network of interest. But as shown in the equation in Fig. 9.1, the large volume (V) of each sphere will result in a very large force when one of these large spheres comes closer to the MRI scanner, especially when we consider that the magnetization of such relatively large spheres will gradually increase as they approach the scanner due to the high fringe magnetic field outside and in the vicinity of the scanner and generated by its large superconducting magnet. The pulling force towards the scanner can be so high that it would not be possible to bring such a relatively large ferromagnetic sphere by hand. This is typically done using the large and powerful robot located in the interventional room, as depicted in Fig. 2.1. When inside the tunnel of the scanner, since there is no magnetic gradient, the large spheres can be manipulated safely and easily by hand, although special separators are used to prevent such spheres from coming together due to large dipole field interactions. To retrieve these spheres one by one once the intervention is complete, the robot is used again.

9.3.4 *Fringe Field Navigation*

Fringe field navigation (FFN) is another approach pioneered by our group. The basic principle is relatively simple. It uses the strong fringe field outside the MRI scanner to generate very strong directional magnetic gradients. Since the fringe field is generated by the superconducting magnet of the MRI scanner, extremely high directional magnetic gradients exceeding 2 T/m can be generated inside the body of a human adult. But FFN, like other approaches, has some drawbacks. First, since it operates outside the tunnel of the scanner, the field strength is lower than the $\mathbf{B}_0$ field. But it is still quite high by being in the range of about 0.5 to 1.5 T, sufficient to achieve high magnetization of superparamagnetic nanoparticles. But the major drawback in terms of magnetic actuation is that the time required to change the direction of our submarines can be relatively long.

Indeed, since moving the bulky MRI scanner containing the superconducting magnet is not practically feasible, instead we robotically move the patient (or animal) in the fringe field surrounding the scanner. This is achieved by connecting the MRI-compatible patient's table to the end effector of the robot located in the interventional room sufficiently far from the high magnetic field generated by the scanner (Fig. 2.1). Depending upon the direction of the gradient that is required to steer the miniature submarines in the bloodstream, the navigation computer calculates the optimal position and angle to position the patient within the fringe field, while providing the maximum possible field strength. This explains the fact that sufficient space in front of the scanner has been reserved for this purpose (Fig. 2.1). Since MRN relies on modifying the ratio of electrical currents in the MRI coils, results in fast directional changes. For DFN, the directional gradients created by the distortion of the uniform magnetic field in the tunnel of the MRI scanner, also results in very fast directional changes during navigation. But since FFN relies on a mechanical motion of the patient table, directional changes are much slower compared to MRN and DFN.

Because the direction of FFN is relatively slowed and the fact that it provides the highest magnetic pulling force, it is more suitable for pulling and navigating thin guidewires prior to the insertion of the injection catheter, as discussed previously in Section 6.1.2.

10

Navigating in the Bloodstream

I N THIS CHAPTER, we are taking a better look at the navigation of the submarine *Proteus* in the bloodstream and comparing it to the navigation methods used with our microscopic submarines.

10.1 Navigating in Larger Blood Vessels (Scene 00:54:00–00:54:14)

Again at the beginning of this scene, we see *Proteus* navigating with the blood flow before steering towards the right artery at another vessel bifurcation prior to proceeding through more bifurcations. Magnetic resonance navigation (MRN), as described in the preceding chapter, is the main technique used to navigate such a microscopic submarine along a preplanned path in the arterial network. But to control the displacement of the submarine, we do not steer it using a joystick operated by a human, as the movie suggests, but rather navigate in autopilot mode, as mentioned earlier in this book.

The reason for using automatic navigation control is simple. The main reason is that each of our microscopic submarines goes too fast to let a human operator to take control. To give an idea, presently what is taking a large part of the total time of the movie showing *Proteus*

A Microscopic Submarine in My Blood: Science Based on Fantastic Voyage
Sylvain Martel
Copyright © 2016 Pan Stanford Publishing Pte. Ltd.
ISBN 978-981-4745-78-9 (Hardcover), 978-981-4745-79-6 (eBook)
www.panstanford.com

navigating from the injection site to the brain typically requires less than one second with our microscopic submarines, a speed well beyond what a human pilot could handle.

So the process of making all the decisions during the navigation phase is done by the navigation computer located in the control room, depicted in Fig. 2.1 and which is located next to the imaging computer display. There are two main categories of navigation control, namely open-loop and closed-loop control. The main difference between open-loop control and closed-loop control, or servo-control in our application, is that unlike for open-loop control, in closed-loop control the last tracked position of the submarine is taken into account (feedback) to make or calculate the next steering decision to maintain the submarine along the planned trajectory. A simple schematic of a closed-loop control scheme is depicted in Fig. 10.1.

The servo-controller depicted in Fig. 10.1 works similarly to the cruise control in your car to maintain a preset speed. First in our particular case, the planned trajectory or command represented by r in Fig. 10.1 is entered in the navigation computer acting as the controller. On the basis of this information, the controller computes an action denoted by u intended to control both the velocity of the microscopic submarine and its direction by applying the proper directional steering force in order to keep the submarine along the preplanned path in

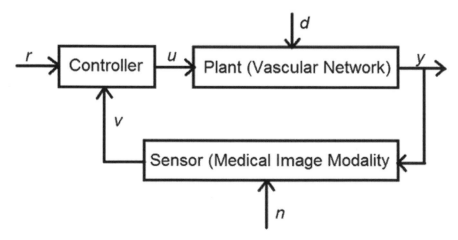

Figure 10.1 Diagram showing the sequence of the various functions involved during the most basic closed-loop MRN.

the blood vessels. This calculated action, u, is sent to the computer-controlled balloon catheter (Fig. 6.3) for flow velocity control and to the MRI gradient coils, through the imaging and navigation electronic cabinets (such as the ones depicted in Fig. 2.4) and located in the electronic and mechanical room (Fig. 2.2) next to the interventional room, for steering (directional) control. All this is referred to as the plant in the schematics, which also includes the vascular network in the patient. During navigation there will be disturbances (denoted by d in Fig. 10.1) that will introduce errors during navigation, which will cause a deviation of the submarine from its planned course. The real position of the submarine submitted to such disturbances and denoted by y in the schematic will be measured (gathered) by a sensor capable of localizing the submarine when operating in the bloodstream. Such a sensor must be a suitable medical image modality, here being a clinical MRI scanner. Such medical imaging modality has a limited resolution and other nonideal characteristics that will introduce errors in the location of the submarine. This is identified as n in the diagram, which stands for "noises." The tracked location (v) acquired with the MRI scanner in this particular implementation is fed back to the controller. On the basis of this localized positon of the submarine, the controller (navigation computer) computes a new decision in the form of new commands to control the velocity and direction of the submarine to maintain it along the preplanned trajectory (path) in the bloodstream. This sequence is repeated at a rate sufficient to maintain the submarine within an acceptable error margin to guarantee that it will reach the targeted location. The basic principle of such a closed-loop MRN sequence is illustrated schematically in Fig. 10.2.

Figure 10.3 shows a more detailed description of the MRN sequence depicted in Fig. 10.2.

There exist various types of control algorithms that can be used by the navigation computer to maintain the microscopic submarines along the planned trajectory. Some of them are listed in Fig. 10.4, along with some of the main issues that must be taken into account when selecting the right control algorithm. The proportional integral derivative (PID) controller is one that is widely used in the industry. It is also the one that was used during our initial navigation experiments performed in the carotid artery using the large 1.5 mm chrome steel

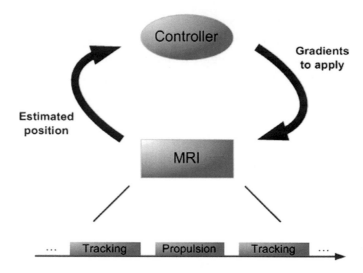

Figure 10.2 Fundamental closed-loop MRN sequence where a controller computes the next *x*, *y*, and *z* gradients generated by the MRI scanner in order to propel (propulsion phase) the microscopic submarine along the planned trajectory based on the MRI-based tracking location (tracking phase).

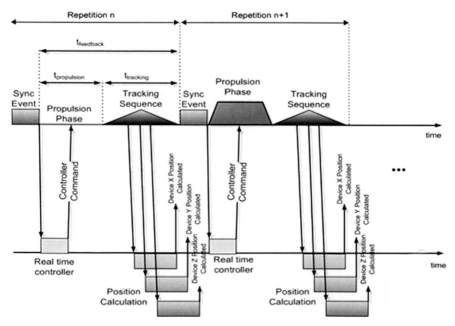

Figure 10.3 Closed-loop MRN sequence that was used during the first navigation of an untethered object in the blood vessel of a living animal. (From *Magnetic Resonance in Medicine* 59:1287–1297, copyright © 2008, with permission from John Wiley and Sons.)

- Some examples of navigation control methods:
    - PID (was used for the first MRN in vivo proof of concept)
    - Self-adaptive fuzzy controller
    - Predictive-based approaches
    - Nonlinear modeling of robust controller-observer
    - Adaptive back-stepping control
    - Etc...
- Main Issues to be considered when selecting the right control method:
    - Increased latency to gather MRI-based feedback positional data
    - Time-multiplexed actuation/gathering positional data
    - Limit in spatial resolution
    - Slew rate associated with the steering gradients, especially high gradients
    - Physiological conditions including blood flow rate
    - Limitations due to the characteristics of the navigable agents
    - Temperature of the coils
    - Limit for MRN actuation duty cycle
    - Etc...

Figure 10.4 Some examples of navigation control methods, along with some of the main issues to consider for selecting the right one.

submarine navigating at 10 cm per second. The closed-loop MRN sequence shown in Fig. 10.3 was done 20 to 30 times per second. To achieve such a high feedback rate, a special MRI-based tracking sequence was also developed by our group.

The PID controller performed well in general, but the submarine was not going in a very straight line all the time. To improve navigation control, mathematical models can be included to help the controller make better decisions by providing more information. One major problem that appears with MRN technology is that the imaging gradient coils of the MRI scanner are used for both tracking the position of the submarine and propelling or steering it. So every time that we need to gather the position of the submarine, the propulsion or steering needs to be shut down. In other words, the nanoengines are still running at full capacity (regime) but the clutch or transmission has to be put in the neutral position to make the submarine visible to the MRI scanner. Since we need all the power that we can get to be able to navigate effectively in the bloodstream, especially when smaller submarines with less superparamagnetic nanoengines (nanoparticles) are used, the obvious solution is to reduce the number of tracking phases such that we can allocate more time for steering the submarines towards the final destination. This is when predictive control can be considered. Indeed, a predictive controller tries to predict the position of the submarines on the basis of several factors, including but not limited to the blood velocity and steering characteristics of the submarine, and to compute the right action to be taken on the basis of these predictions. It is somewhat like in the scene beginning at 00:46:25, where General Carter is trying to predict how long *Proteus* will need to navigate across the heart using a scientific ruler, except that in our case, a powerful computer is used instead.

But there are uncertainties such as turbulences and other factors that are hard to predict and to model mathematically, which may affect the efficacy of a predictive controller. Measuring the actual blood flow fast enough and with sufficient accuracy is technically very challenging. The use of other systems such as an optical fiber connected to an OCT system (see Fig. 6.3) is one potential option. But there are measurement errors associated with it, and it can only measure very near the tip of the fiber. When navigating deeper in the vascular network, these data become less reliable. Then at some point, when predictive control is trying to predict too far ahead, it becomes as reliable as predicting the weather two weeks in advance.

To relax the tracking specifications, we could decide to not really care about the exact position of the submarine but only allow the controller to take action on the basis of an approximate location of the submarine. These approximate locations would be linked to a predefined set of rules or actions. Depending on the regions in which the submarine would be located, a particular action linked to this region would be initiated. Such an approach is referred to as fuzzy control, and it is illustrated in Fig. 10.5.

For a fuzzy navigation controller, for instance, when the submarine is in a specific region, this could trigger a specific preset action such that the generation of a maximum magnetic gradient oriented in a determined direction to maximize the efficacy of the submarine to take the right-branching artery at maximum velocity.

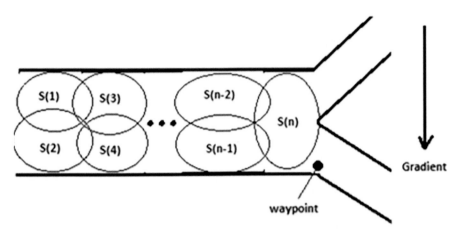

Figure 10.5 An example of regions requiring a specific action for a fuzzy controller to take the right-branching artery.

These are just some examples of possible closed-loop control algorithms that can be considered for navigating submarines in the bloodstream. An open-loop navigation control approach can also be used to allow maximum time for applying the steering gradients. In this case, a few submarines that we called "scouts" are navigated, following a predetermined set of gradient or propulsion sequences. Then an MRI of the submarines is taken to detect their final position. This phase is repeated by correcting the propulsion sequence until it proves to be effective. Then the rest of the submarines are injected and submitted to the same propulsion sequence. An example is depicted in Fig. 10.6.

One of the problems with closed-loop and open-loop MRN sequences when there are several changes in the direction of travel of the submarine during the trip, especially when there are large numbers of submarines involved, is the overheating of the gradient coils. By the way, *Proteus* had also an overheating problem. Check it out in the scene 01:19:20–01:1923.

Indeed, for MRN, the steering gradient generated at a particular time will apply to all microscopic submarines navigating in the bloodstream. As such, the number of microscopic submarines that can be navigated at the same time would decrease as we travel deeper in the vascular network where the length of the blood vessels separating

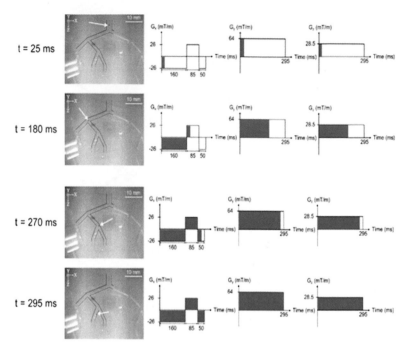

Figure 10.6 An example of an open-loop navigation (propulsion/steering) sequence. From *IEEE Trans. Robotics*, 30(3):719–727, copyright © 2014, with permission from IEEE.

two successive bifurcations will become shorter. This means that there will be more injections and therefore more repeated navigation phases that would lead to more directional changes and therefore more overheating of the gradient coils. Such overheating will force the navigation sequences to be temporarily stopped until the coils cooled down. In turn, this may extend the time required to complete the intervention, which might not be the ideal scenario for a hospital.

One possibility is to apply several gradients pointing to different directions at the same time along the planned trajectory. This is exactly what dipole field navigation (DFN) explained in the preceding chapter is doing. The advantage is that all microscopic submarines can be navigated simultaneously, therefore requiring only one injection. There is no overheating since the gradients are generated by large ferromagnetic spheres and not by the coils. But since the large magnetic dipoles generated by these large spheres distort the $\mathbf{B}_0$ field, proper MRI and magnetic resonance tracking cannot be done, forcing this approach to be an open-loop navigation method.

10.2   Navigating in Narrower Blood Vessels (Scene 00:54:14–00:54:42)

In this scene, *Proteus* is starting to navigate in narrower blood vessels. As transiting to such narrower blood vessels does not seem to affect too much their navigation capability, it becomes much more challenging for us. Open-loop navigation would become much harder and unpredictable. The latency or delay in gathering the tracking position of the submarines and applying a directional gradient within a shorter distance between successive bifurcations would require slowing down the blood velocity, which might not be possible at such a relatively long distance from the balloon catheter. Obtaining a map of narrower blood vessels would be more difficult and often not possible due to the limited spatial resolution of existing medical imaging modalities. Smaller submarines capable of entering such narrower blood vessels would emit weaker signals (smaller image artifacts) that would require different MRI sequences that most likely would need more time to complete, etc. These are just some of the challenges. Nonetheless, by pushing the technologies, the submarines are still able to navigate close enough to the capillaries to allow our crew members to swim towards the target.

11

The Crew

W E FINALLY PRESENT in this chapter our microscopic crew members. As you will see, they may be more qualified for such a mission than the crew members of *Proteus*.

11.1 Low in Oxygen (Scene 00:56:00–01:06:56)

In this long scene, you witness the crew members working very hard to get the oxygen that they lost following a malfunction of a valve in the submarine *Proteus*. The small amount of air remaining in the tank was insufficient to continue the mission. It is too bad that *Proteus* did not have our crew members, because unlike the crew in the movie, our crew needs very little oxygen to survive. In fact our crew members are happy with only 0.5% oxygen.

11.2 In the Human Mind (Scene 01:27:11–01:28:26)

In this scene, the crew of the submarine *Proteus* is entering the brain. Unfortunately, entering the brain is not as easy as it looks in the movie. This is because of the blood–brain barrier (BBB), which has tight

A Microscopic Submarine in My Blood: Science Based on Fantastic Voyage
Sylvain Martel
Copyright © 2016 Pan Stanford Publishing Pte. Ltd.
ISBN 978-981-4745-78-9 (Hardcover), 978-981-4745-79-6 (eBook)
www.panstanford.com

junctions preventing an object, even as small as the crew members of *Proteus*, from being able to enter. A more probable scenario here would have been the use of what is known as high-intensity focus ultrasound (HIFU), which has proven to be able to open the BBB to allow larger objects to enter but certainly not at the size of *Proteus*. But HIFU did not exist at the time of the movie.

What is also interesting in our particular case is that HIFU has been and can be used effectively in a clinical MRI scanner. But the propagation of sound is affected by tissue density, and the difference in density between bones and soft tissues, including the brain, is quite significant. The result is that is it very difficult to open the BBB exactly where the submarine would be.

In our particular case, we use the same superparamagnetic nanoparticles used to propel our microscopic submarine and as a tracking device, allowing the position of the submarines to be tracked by MRI, to open temporarily the BBB. This is achieved using the same principle used to miniaturize the hydrogel-based microscopic submarines (Chapter 5) using localized hyperthermia based on the Néel effect. The main advantages are that magnetic fields are not affected by differences in tissue densities. Since the opening is triggered by the nanoparticles. Therefore, the opening will be done exactly at the location of the submarines. For this to happen, the patient needs to be transferred to the hyperthermia station using the robotic table (Fig. 2.1). At the end of the scene, the crew sees the clot. In our case this would most likely be a solid tumor.

11.3 Crew Members Exiting the Submarine (Scene 01:30:06–01:30:15)

In this scene, the first crew member exits the submarine by opening a hatch. Various mechanisms can be used for our crew members to exit the microscopic submarine. These methods include but are not limited to a biodegradation of the hatch within a relatively short time; an automatic opening of the hatch, triggered by a change of pH around the tumor; and a computer-triggered opening using the shrinkable process of the hydrogel submarines, as described in Chapter 5.

11.4 Crew Members Swimming towards the Target (Scene 01:30:48–01:31:07)

In this scene, the crew members of *Proteus* swim towards their target. This scene is similar to our crew members swimming towards the tumor, as depicted in Fig. 6.2. If you look more closely, you will see what a crew member looks like (Fig. 11.1).

Each of our crew members is indeed a bacterium but a special one. It is known as the MC-1. Instead of two legs like the divers in the movie, our crew members have two flagella bundles identified in Fig. 11.2 as a propulsion system. Indeed, these flagella propels the MC-1 at an average velocity of 200 body lengths per second, 200 times the swimming speed of the best Olympic swimmers (in body lengths per second). It is so efficient because flagella have been proven to be extremely effective in low Reynolds number conditions unlike the legs for the crew members of *Proteus*. Our swimmers are even smaller than a red blood cell, as shown in Fig. 2.2.

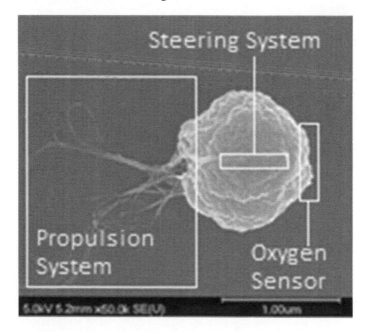

Figure 11.1 Microscopic image of one of our crew members that we believe are more qualified than the crew members of *Proteus* for such a mission.

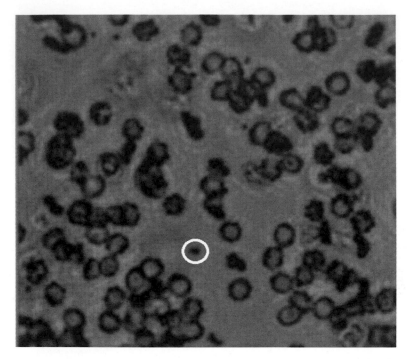

Figure 11.2 One of our microscopic swimmers (bacterium) being shown in the white circle next to human red blood cells.

Now the question is how to tell or communicate to these swimmers in which direction they should swim to get to the tumor. Well, the secret is inside the bacterium itself (identified in Fig. 11.2 as the steering system) in the form of a chain of iron oxide nanoparticles. Such a chain is in fact a nanomagnetic compass needle. Indeed, it works exactly like a magnetic compass needle that point to the North Pole. These types of microorganisms are known as magnetotactic bacteria.

To indicate the direction of the tumor (or a clot as in the movie), we simply generate an artificial North Pole at the location of the tumor. While the bacteria swim, a directional torque induced on such a chain of nanoparticles from a very weak magnetic field will orient the bacteria towards where we want them to go, and since we can see the tumor mass by MRI, we know exactly in which direction they should swim. The magnetic field used to guide the bacteria towards the tumor in our case is done by a special platform dubbed the "magnetotaxis system." Such a system is capable of generating a magnetic North Pole in a 3D

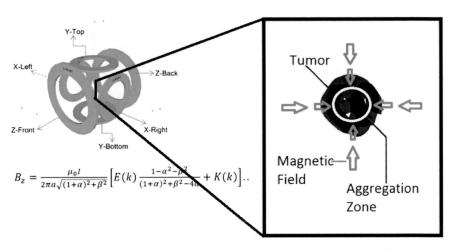

$$B_z = \frac{\mu_0 I}{2\pi a\sqrt{(1+\alpha)^2+\beta^2}}\left[E(k)\frac{1-\alpha^2-\beta^2}{(1+\alpha)^2+\beta^2-4\alpha}+K(k)\right]..$$

Figure 11.3 One version of the magnetotaxis system showing the generation of an artificial North Pole where millions of bacteria aggregate (3D aggregation zone); the overall size of the aggregation zone is determined by the amount of electrical current circulating in the coils (upper-left corner). Outside the aggregation zone, the directional magnetic field is sufficiently high to influence the direction of the magnetotactic bacteria (magnetotaxis). When inside the aggregation zone, the magnetic field is no longer strong enough the influence the swimming direction of the bacteria. It is inside the aggregation zone that the bacteria switch to aerotaxis, seeking regions of low oxygen concentrations that correspond to hypoxic regions in tumors. These hypoxic regions are the targeted zones that must be destroyed.

volume anywhere in the human body. As such, when the submarines have reached their final destination and cannot go further, the patient would be transferred from the MRI scanner to the magnetotaxis system (see the plan of the interventional room in Fig. 2.1). This is depicted in Fig. 11.3.

Following the direction of the magnetic field in the magnetotaxis system, the bacteria will aggregate in the artificial North Pole (aggregation zone in Fig. 11.3).

11.5 Armed Crew Member (Scene 01:31:16–01:31:30)

In this scene, we see one of the crew members being armed with a laser gun. In our particular case, since we do not have miniature laser guns, the crew members carry chemical weapons instead, as depicted in Fig. 11.4.

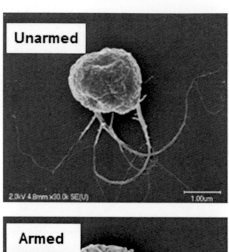

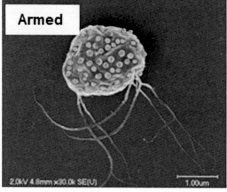

Figure 11.4 This is one example of an armed crew member being represented by a magnetotactic bacterial cell loaded with 70 small nanocontainers (known as liposomes), each containing the chemical molecules that will kill the cancer cells.

11.6 Destroying the Target by Pointing and Shooting the Laser to the Right Locations (Scene 01:31:30–01:32:32)

In this scene, the laser is pointing at locations that will destroy the clot. A well-trained surgeon who knows where to shoot is manipulating and pointing the laser in the right directions. In our case, the regions where our attack will result in the maximum killing effect on the tumor are known as hypoxic regions. But since we do not have miniature surgeons that can identify these hypoxic regions within the tumoral mass and as they cannot be visualized by any external medical imaging modality, how do we train these armed bacteria to release their chemical charges

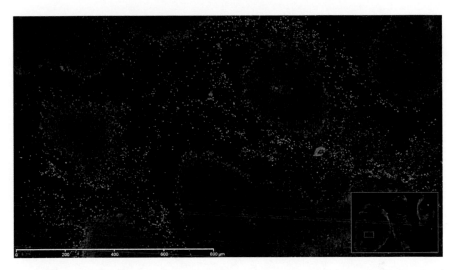

Figure 11.5 Image of the inside of a tumor, showing armed bacteria (green dots) in hypoxic zones.

in these hypoxic regions? Well, these bacteria are microaerophilic, meaning, as mentioned earlier, that they prefer environments of low oxygen concentration, more precisely, an oxygen concentration of 0.5%, which by coincidence corresponds to the oxygen concentration found in hypoxic regions of tumors.

So, when in the aggregation zone, since the magnetic field is not sufficiently strong to guide them, our microscopic crew members do not know what to do, and naturally, they follow the decreasing oxygen gradient until they reach the regions with 0.5% concentration of oxygen. This is shown in Fig. 11.5 depicting these armed bacteria (green dots) in the hypoxic regions of a tumor.

11.7 *Proteus* Being Attacked by the Body Defense System (Scene 01:34:09–01:35:47)

In this scene, *Proteus* is being attacked by the body defense system. We witnessed a similar situation with the nanoengines of disintegrated microscopic submarines in the brain of a rat. This is depicted in Fig. 11.6.

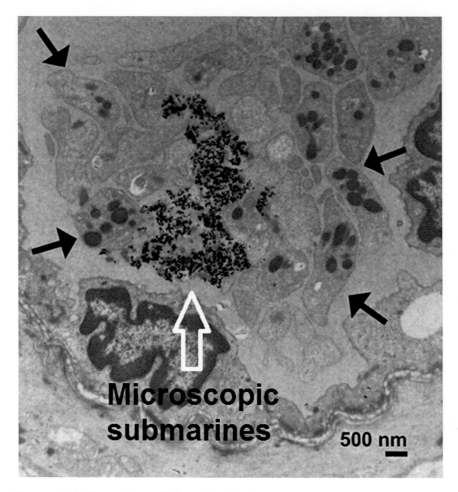

Microscopic submarines

500 nm

Figure 11.6 Macrophages indicated by black arrows are approaching the super-paramagnetic nanoparticles of the disintegrated microscopic submarines (indicated by the white arrow). Adapted from *Journal of Controlled Release*, 206:49–57, copyright © 2015, with permission from Elsevier.

As for the remaining of the movie, we see how the crew members escape alive. For our crew members, their destiny is to die and most likely being flushed, at some point, down a toilet. As for the rest of the microscopic submarine, just look at Fig. 11.6.

Conclusion

I N THIS BOOK we saw how the scenario of the movie *Fantastic Voyage* written 50 years ago could be mimicked using the right ingredients—the use of a multidisciplinary approach mixing various engineering and scientific disciplines with a bit of imagination. We also showed that harnessing what nature has already provided and nanotechnology were two critical components that made such a scenario possible.

This effort will enable a first new interventional room that will, I hope, help save a lot of lifes. Until then, our fantastic voyage will continue.

Index

actuate 107
adventure 27
agent, security 14
aggregations 84, 86
  large 46, 75
aggregation zone 127, 129
airport 6–7
animals, living 30, 32–33, 73, 99–100, 116
arrow 20, 46–47, 90, 103
arterial network 30, 57–58, 61, 74, 90–91, 110–111, 113
arteries 17–18, 26–27, 30, 32–34, 45, 60, 73, 76, 81, 83, 85, 88–90, 103, 105
  right-branching 118–119
assembly 40–41, 92, 107
atoms 47, 49, 95–96
axis 44–45, 94–96, 98
  horizontal 44–45
  vertical 44–45

bacteria 86, 126–127, 129
  armed 128–129
  magnetotactic 126–127
bacterium 125–126
balloon catheter 33, 85, 121
BBB, see blood–brain barriers
bifurcation
  next 84–85
  successive 74, 120–121
blockage 75
blood 17, 25, 37, 41, 49, 57, 69, 72, 74, 81, 86–87, 101, 106, 113, 123

blood–brain barriers (BBB) 19, 34, 123
blood flow 33, 51–52, 85, 101, 103–105, 113, 118
blood flow rate 33, 73
bloodstream 18–19, 21, 23–24, 29–30, 32, 34–35, 38–39, 45, 48–49, 51–52, 54, 110, 112–113, 115, 118–119
blood velocity 118, 121
blood vessels 30–32, 34, 37, 51–52, 54, 72–73, 76, 78–79, 86, 88, 90, 99–100, 107, 109, 115–116
board 12, 29
  white 12–13
body defense system 129
body lengths 70, 72, 83–85, 100, 125
body temperature 61, 63
bottles 64–65, 67
brain 20, 26, 28, 30, 34, 45, 54, 60, 72, 76, 114, 123–124, 129
building 5, 7–8, 10–11, 14–15, 30
building submarines 82

cabinets 19, 23
campus 6, 11–12
cancer 3, 27, 54, 78
carotid artery 30, 32–34, 37–38, 42, 49, 53, 55, 57, 65, 72, 79, 81, 83, 100, 103
catheter 74, 91
cells, red 81
chain 40, 126
chaotic 12–13

chemoembolization 76
circles, white 39, 61, 90–91, 126
clinical MRI scanner 32, 51–52, 78, 88,
        94, 96–97, 103, 107–110, 115,
        124
clinical MRI scanners, modern 51
closed-loop MRN sequence 115–117
clot 26, 54, 75, 124, 126, 128
cobalt 44, 49
coffee 92
coils 43, 65, 107, 109, 120, 127
    electromagnetic 98, 107
commands 26, 29, 31, 114
computed tomography (CT) 94
computer 12, 28, 30, 32, 75, 97, 118
computer display 29, 31, 52–54, 100
computer-generated image 89–90
computer sciences (CS) 28
computer screen 52, 54
concentrations 62, 129
    low oxygen 127, 129
conductor 28–29
configuration 43, 73
construction 39–40
control 2, 73, 75, 77–78, 113–115
    closed-loop 114
    open-loop 114
    predictive 118
control computer 32–33
controller 114–118
    predictive 118
control room 19–20, 37, 39, 52, 95,
        114
cost 43, 50
crew 35, 39, 41, 57, 63–64, 66–67, 70,
        75, 77, 123–124
crewed submarines 24, 39–40, 60–61,
        72, 76
crewed submarines exit 41
crew members 19, 23, 25–26, 29,
        34–35, 37, 41, 63–65, 70–71, 81,
        83, 121, 123–125, 127, 130
    armed 75–77, 127

miniature 26, 39
    population of microscopic 67
crew members onboard 39–40
crew members swimming 125–127
cross-linkers 62–63
cryostat 43
CS, see computer sciences
CT, see computed tomography
CT scanner 96

danger 35
depicts 9, 11, 30, 32, 37, 39–40, 44, 59,
        73, 82, 89, 100, 105
depth 51–52
DFN, see dipole field navigation
diagram 59, 61, 114–115
diameter 38, 42, 48–49, 60, 71–73,
        76–77, 82, 89, 91, 97, 103, 109
    inner 107–110
difference 9, 11, 17, 27, 30, 42, 82, 124
dimensions 22, 39, 58, 81, 105
dipolar 78, 84
dipolar interactions 46, 83–84
dipole field navigation (DFN) 85, 109,
        111–112, 120
direction 47–49, 51–52, 60, 72, 84, 90,
        95, 98, 101, 103, 106–109, 112,
        114–115, 119, 126–127
    right 84, 90
    swimming 76, 127
directional 114–115
directional changes 73, 112, 120
    fast 112
directional gradients 78, 109–110, 112,
        121
directional magnetic force 107
directional torque 106, 126
dishes 63, 65–67
display 54, 91, 99–100
distance 51, 70, 78, 84–85, 98, 103–104,
        121
distortion 109, 112

disturbances 115
doctors 49, 75, 89, 91
doors 8–10
    large 9
dots
    green 90, 129
    small 39–40, 82
drawbacks 112

efficacy 18, 118
elapsed time 74
electromagnetic units 65–67
electronic cabinets 19, 23–24, 115
elevator 7–8, 11
embolization 76
engineers 2, 18, 22, 28
engines 42–43, 45, 49, 106–107
    miniature microjet 106
    smaller 102
entrance 8–9, 14–16, 21, 41, 78, 94
    main 14–16
environments 28, 129
equation 102–103, 107, 111
errors 115
exit 9, 25, 61, 63, 74–75, 124
experiments 26, 34

fabrication 41
facility 6–7, 11–12, 14–17, 19–21, 23,
    37, 43
ferromagnetic 42, 46, 49
ferromagnetic core 65
ferromagnetic materials 42, 44, 48
ferromagnetic particles 42–49, 102
ferromagnetic spheres 109–110
    large 110–111, 120
FFN, *see* fringe field navigation
fiber, optical 74, 118
field 22, 28, 43, 50–52, 67, 76, 78, 84,
    94–96, 98, 103, 109, 112, 120
field strength 103, 110, 112

flagella 125
floor
    main 11, 14
    next upper 9, 11
flow 85–86, 101, 104–105
    pulsatile 104–105
fluctuate 48
fluid 41, 69, 75
    quantity of 64–65
force 45, 48, 51, 75, 78, 102–103, 120
    interactive 84
    pulling 89, 111–112
formation 41, 46, 66, 75, 84
frequency 60, 92
fringe field 78, 112
fringe field navigation (FFN) 112

goal 2, 8, 15, 26, 38
gradient coils 99, 107–108, 119–120
gradient field 98
gradients 78, 98, 107–109, 112, 116,
    119–120
    steering 119
graph 44–45, 63, 78
gravity 100
grid 95
group 29, 57, 73, 85, 88, 107, 112,
    117
guide 88, 91, 126, 129
guidewire 88–91

hatch 41, 124
head 27, 92, 107, 109
    patient's 92–93
heart 29, 34, 101, 104–105, 118
heat 43, 60
hepatic artery 37–38, 49, 72–73, 76,
    97–98, 104
hexagons 63
HIFU, *see* high-intensity focus
    ultrasound

high-intensity focus ultrasound (HIFU) 124
high-magnitude directional $\mathbf{B}_0$ field 95
holes 73, 85
Hollywood movie 22
horse power 49
hospitals 8, 19, 29, 51, 120
human body 52, 58, 60, 76, 87, 95, 109, 127
human hair 22–23, 39, 77, 96–97
humans 18, 27, 38, 70, 72
humor 35
hydrogel submarines 62, 124
hydrogen atoms 23, 95–96
hydrophilic 40–41, 58
hyperthermia 33–34, 59
hyperthermia station 60, 124
hypothermia 34

IGC, *see* imaging gradient coil
image artifact 90–91, 96
images 15, 17, 30–33, 53–54, 58, 77, 89, 91–93, 97, 99–100, 108, 110–111, 129
   projected 88–89
   real 30, 53, 89–90
image slices 98, 107
imagination 3, 131
imaging 19, 23, 51, 54, 94, 108–109, 115
imaging gradient coil (IGC) 107, 109, 117
infrastructure 5, 8
inject 26, 35, 64–65, 69, 72–75, 78, 85, 91
injection 18, 63–64, 67, 69–72, 74–75, 78, 81, 85, 91, 120
   peritumoral 70–72
injection process 69, 75
injection site 2, 69–70, 114
injection system 73–75, 78
injector 75, 78

inlet 74
instruments 28–29, 42
interventional facility 5, 8, 10, 12, 14–17, 19, 21, 37
   medical 16
   new 2, 6, 11, 18
interventional room 19–20, 51–52, 60, 88, 94–95, 111–112, 115, 127
interventions 18, 27, 29–30, 38, 52, 87, 111, 120
invasion 77
iron 44, 49
iron–cobalt 48–49
iron oxide 48, 103

joystick 54, 113

laser 54, 75, 128
laws 22
LCST, *see* lower critical solution temperature
legs 125
length 81, 83, 119
letters 21
life 26–27, 34, 131
lifetime 63
liquid 64–65
liquid helium 43
liver 28, 37–38, 49, 72, 76, 97–98, 104
location 5, 7, 17, 19, 27, 39, 41, 44, 94, 96, 98, 100, 124, 126, 128
   approximate 118
   targeted 70, 72, 89–91, 115
logo 21
lower critical solution temperature (LCST) 58–60, 63

magnet, permanent 42–44, 46
magnetic 42, 44–45, 47, 49, 60, 65, 89, 91, 102

magnetic domains 46–48

magnetic field 42–48, 50–52, 60, 78, 95–96, 98, 102, 106–107, 124, 126–127, 129
   applied 50, 59–60, 90, 106
   applied directional 47
   dipole 78, 110
   directional 47, 127
   external 43–44, 47, 60
   high 49, 51, 112
   weak directional 106

magnetic field strength 44, 103–104, 107

magnetic gradient 51, 98, 103–104, 107, 109, 111
   directional 51, 107
   high directional 112
   strong directional 112

magnetic guidewire 91

magnetic nanoparticles 44, 58–59, 61, 78

magnetic particles 93, 96

magnetic resonance imaging (MRI) 8, 19, 35, 49, 88, 90, 92–94, 96, 98, 100, 107, 119, 124, 126

magnetic resonance navigation (MRN) 73, 85, 106–107, 112–113, 119

magnetic resonance submarine 55

magnetic submarines 91, 94, 100
   hydrogel-based 58
   microscopic 94
   single 97
   small 104

magnetic tip 89–90

magnetization 42–43, 45, 47–49, 76, 78, 102–103, 111
   induced 44–45
   residual 43, 46, 48

magnetization curve 44, 46–47

magnetization level, saturation 45

magnetotaxis system 126–127

magnitude 42, 44, 50–51, 95, 98

map, large 87–88

materials 38–39, 46, 58

maximum regime 48

medical imaging modalities, best 93–94

medicine 32

meeting 30, 35, 37

meeting room 13, 29–30

members, microscopic 67

meshes 95

meters 70, 83, 85, 98

metric system 82

microcatheter 73, 75, 91

microjets 105–106

microscopic crew members 39, 59, 61, 63–67, 69–70, 72–73, 76–77, 82, 104, 123, 129

microscopic submarines 69, 73–78, 81, 84–85, 87, 91, 93–94, 96–98, 101–102, 104–105, 107–110, 113–116, 119–120, 123–124, 130

microscopic submarines approach 78

microscopic submarines navigating 119

microscopic swimmers 72–73, 126

millimeter 54, 70

mimics 2

mind 2

miniature magnetic submarines 96

miniature submarines 17, 33, 35, 41, 45–46, 51, 72, 87, 112
   jet-propelled 105

miniaturization 57, 59–60, 62, 103
   level of 61–62, 94

miniaturization phases 59, 63

miniaturization process 57, 60–61, 63, 65

miniaturize 34, 54, 60, 63, 65, 124

mission 20, 25–26, 34–35, 45, 64–65, 67, 75–77, 123, 125
   complex 25, 27
   first 18
   suicide 26

modulating 59–60, 105

molecules 40, 49, 62
   chemotherapeutic 76

movie 5–9, 19–21, 25–27, 29–30,
    33–35, 37–38, 41–42, 59–61,
    63–65, 71–73, 75–76, 87, 91–93,
    104–106, 123–126
movie claims 92
movie *Fantastic Voyage* 2, 5, 22, 30, 64,
    82, 131
MRI, *see* magnetic resonance imaging
MRI machine 51
MR-imaging 54
MRI scanner 51–52, 76, 78, 84, 88–89,
    93, 95–96, 98, 100, 107, 109–112,
    115, 117, 127
MRN, *see* magnetic resonance navigation
music 29
musicians 29

name 2, 19, 21, 54–55, 64
nanoengines 45–46, 49, 51, 78, 84, 103,
    107, 109, 117, 129
nanometers 21, 23, 48–49, 54–55, 60
nanoparticles 45, 48–49, 51, 60, 102,
    118, 124, 126
nanotechnology 22–23, 46, 55, 60,
    131
narrower blood vessels 38–39, 76, 89,
    104, 121
naval officer 29
navigate 2, 32–34, 38, 42, 45, 49,
    51–52, 57, 61, 73, 76, 103–104,
    113, 117–118, 121
navigation 19, 23–24, 27, 33–34, 38,
    51–52, 54, 74, 85, 88, 100, 106,
    112–113, 115
  open-loop 120–121
navigation computer 54, 74, 97, 112,
    114–115
navigation control 23, 114, 117
  automatic 32, 113
navigation phase 18, 32, 53, 114
navy 29, 74, 83, 85
needle 47, 50, 69, 71

neutral position 51, 107, 117
North Pole 46–47, 50, 126
  artificial 126–127

objects 22, 25, 35, 44, 76, 124
  ferromagnetic 44–45
OCT, *see* optical coherence tomography
office 13
operation 43, 89, 106
optical coherence tomography (OCT)
    74, 94
option 34–35, 54, 91
orchestra 28–29
overheating 119–120
oxygen 49, 72, 95, 104, 123, 129
oxygen concentration 129

particles 41–42, 45–46, 48, 74, 96
  microscopic 42, 45, 92
path, preplanned 113–114
patient 17–19, 30, 33–34, 46, 49, 51,
    69, 76, 78, 92, 94, 96, 98, 109–110,
    112
payloads 59, 63
penetrate 60, 94
people 6, 8–9, 11, 22, 27, 72, 82
perspective 5, 22–23, 33, 70
PET, *see* positron emission tomography
PET-based tracking 92–93
PET scanner 92–93
PET scanners, first 92
phantom 88
phase 57, 61, 63, 119
photograph 7–10, 12–16, 23–24, 58, 66,
    71, 88–89, 92–93, 105
physics 22
physics book, traditional 22
PID, *see* proportional integral derivative
pig 17–18, 30
pilot 52–53, 101
pilot seat 52

plan 20, 25–26, 34–35, 127
planned trajectory 31–32, 98, 103,
    114–116, 120
platforms 5, 19–20, 46, 59, 63–65,
    75–76
PNIPA hydrogel submarines 58
PNIPA magnetic submarine 59
PNIPA submarines 58, 61–63
    crewless 60
    initial 61–62
    navigable 59
population 65–67, 71
position 29, 32, 34–35, 53, 75, 87,
    90–95, 97, 99–100, 112, 117–118,
    124
    exact 99, 118
    tracked 88, 114
positron emission tomography (PET)
    35, 92–93
power 41–42, 46, 49, 65–66, 76, 86,
    92–93, 117
power supplies 65–66
principles 2, 22–23, 47, 95, 124
process 53, 57, 60–61, 63, 66–67, 74,
    114
project 30, 48, 50
propellers 105
propelling forces 38, 45, 48, 103–104
    available 101
    total 101–102
proportional integral derivative (PID)
    115
propulsion 42, 101, 103, 105–106, 117
propulsion sequences 119
propulsive force 102
Proteus being attacked 129–130
Proteus navigating 83, 113
Proteus swim 125
protons 95

radars, small 92
radiation sensors 92

radio frequency (RF) 95
range 22, 70, 81–82, 84, 112
red blood cells 22, 83, 125
regime 51–52, 78, 107, 109, 117
regions 37–39, 47, 49, 60, 70–72, 76,
    86, 97, 104, 118–119, 127–129
    hypoxic 127–129
relaxation time 95–96, 98
remanence 43–45
research 12, 29, 50
resolution 92, 94
Reynolds number 86, 105
RF, *see* radio frequency
RF pulse 95–96
robot 21, 111–112
role 19, 27–28
rooms 12, 18–21, 91
    mechanical 19, 23–24, 42, 115
    preparation 17, 19
    sterilization 35, 37
route 30–31, 34

saturation magnetization 48–51, 78, 96,
    103, 107, 109 110
scanner 8, 19, 51, 78, 94, 98, 109–112
scenario 2, 19, 49, 59–60, 72, 79, 105,
    124, 131
scene 6–9, 14–17, 19, 23, 25–27, 29–30,
    33–35, 41, 54–55, 75, 81–87,
    91–92, 104–106, 123–125,
    127–129
    first 6, 17
    following 57, 81
    long 83, 123
    next 91
    short 25–26, 35, 82–83, 95, 101, 105
scene beginning 118
school 28
science fiction movie 22
sciences 2, 22, 27, 29
scientists 2, 18, 28, 82
screen 30

screws 62–63

SDS, *see* sodium dodecyl sulfate

section 12–13, 16, 29, 91, 105, 107, 112

security desk 14

sensor 115

sequence 109, 114–115, 120

SGC, *see* steering gradient coil

ship 29, 85, 105

shrinks 26

signal 95–96, 100

size, average 83

skin 69, 71, 104

smoking 30

sodium dodecyl sulfate (SDS) 106

South Pole 46

space 51, 72, 99, 107, 110, 112
 living 41
 parking 7–8

spatial resolution 93, 96, 98–99

specializations 27–29

specialties 28

spheres 39, 46, 109, 111
 large 111, 120

stairs 11
 mechanical 11–12, 23

steering gradient coil (SGC) 107, 109

structure 58, 61–62, 76

submarine navigating 117, 119

submarine *Proteus* 61, 69, 81, 87, 92, 101, 113, 123

submarines 17–19, 23–26, 29–35, 37–43, 45–46, 48–55, 57, 59–63, 69–70, 74–79, 81–87, 96–105, 112–115, 117–119, 124
 crewless 37, 49, 73
 design 35
 disintegrated microscopic 129–130
 first 42, 55, 83
 hydrogel-based microscopic 124
 largest 85
 micrometer 82
 microscale 78
 miniaturized 29
 nanoparticle-powered 44

navigable 40, 58, 94

neighboring 78

next-generation 49

real 57, 61, 105

shrinkable 58

single 97

single-passenger 82

small 94

smaller 45, 81, 100, 103, 118, 121

submarine sizes 81

superconducting magnet 42–43, 51, 78, 112

superparamagnetic nanoengines 48, 102, 105, 118

superparamagnetic nanoparticles 44, 46–48, 51, 60, 76, 78, 102–103, 107, 109–110, 112, 124
 iron oxide 49–50

superparamagnetic submarines 78
 microscopic 107

swim 39, 70, 72–73, 76, 97, 121, 126

swimmers 86, 125–126
 human 70, 72

synthesis 40–41, 103

syringe 63, 65, 67, 69–71, 73, 75, 91
 simple 69, 73

team 18, 25–30, 35, 40, 64, 81, 91

team members 25, 27–30
 good 25, 27

technologies 2, 14, 18, 21, 26, 29, 41, 46, 92, 121
 new 22, 50

temperature 20, 33, 48, 58–59

temperature level 34

therapeutic agents 2, 54

thickness 23, 77, 96, 98

TIFP, *see* tumor interstitial fluid pressure

time 5–6, 22–23, 26, 28, 61–64, 73–75, 78–79, 82–83, 85, 95, 100, 109, 112, 117–121, 124
 first 17, 30, 32–34, 37, 81
 long 50, 109

maximum 25–26, 119
  short 2, 60, 124
  total 63, 113
tip 38, 42, 74–75, 78–79, 89–91, 118
tissue densities 124
top 9, 41, 53–54, 63, 66, 98, 103
torque 47
track 19, 34, 75, 87–88, 91–92, 94
tracking 23–24, 34, 87, 92, 94, 109,
    117
  real-time 54
tracking phases 116, 118
tracking position 53, 90, 99, 121
tracking system 92–93
transit 11, 15, 103–104
transiting 6, 9, 15, 121
tumor 28, 54, 70–73, 76–77, 97, 104,
    124–129
  solid 70, 76, 124
tumor interstitial fluid pressure (TIFP)
    72
tunnel 51–52, 78, 84, 94, 96, 98,
    107–112
turbulences 86, 118

underground floor 9–10, 14
underground level 8–9
uniform $\mathbf{B}_0$ field 107, 109
university 29
university campus 7–8

untethered object 27, 30, 32–33, 38,
    100, 116

vascular network 45, 87–89, 91, 97,
    103, 115, 118–119
velocity 73–75, 83, 85, 100, 105,
    114–115
  high 75, 78, 103
volume 51, 65–66, 77, 96, 103, 127
  large 64, 66, 111
  total 102

walls 21, 30, 88–90, 95, 105
  arterial 85–86, 89
  external 39–41
  submarine's 41
war 77–78
water 58, 67, 72, 86, 95
wavelength 94–95
waypoints 31–33, 53
windows 54, 95
  large 12, 95
wires 13, 89
woman 27
  single 27
women 27

x-rays 30, 88, 94, 96